Advances in Aquatic Ecology

— Volume 3 —

Advances in Aquatic Ecology

— Volume 3 —

Editor
Dr. Vishwas B. Sakhare
Head,
Post Graduate Department of Zoology
Yogeshwari Mahavidyalaya,
Ambajogai – 431 517
Maharashtra

2010
DAYA PUBLISHING HOUSE
Delhi - 110 035

ISBN 81-7035-633-4
ISBN 978-81-7035-633-2

Published by : **Daya Publishing House**
1123/74, Deva Ram Park
Tri Nagar, Delhi - 110 035
Phone: 27383999
Fax: (011) 23260116
E-mail: dayabooks@vsnl.com
website: www.dayabooks.com

Showroom : 4760-61/23, Ansari Road, Darya Ganj,
New Delhi - 110 002
Phone: 23245578, 23244987

Laser Typesetting : **Classic Computer Services**
Delhi - 110 035

Printed at : **Chawla Offset Printers**
Delhi - 110 052

PRINTED IN INDIA

Preface

I am delighted to write about the third volume of *Advances in Aquatic Ecology*. This volume is the compilation of esteemed articles of internationally acknowledged experts in the field of aquatic ecology with the intention of providing a sufficient depth of the subject to satisfy the need of a level which will be comprehensive and interesting. It is an assemblage of up to date information of rapid advances and developments taking place in the field of aquatic ecology. With its application oriented and interdisciplinary approach, I hope that the students, teachers, researchers, scientists, policy makers and environmental lawyers in India and abroad will find this volume much more useful. The articles in the book have been contributed by eminent scientists/academicians active in the areas of aquatic ecology.

My special thanks and appreciation go to the scientists whose contributions have enriched this volume. I wish to express my sincere gratitude to Dr. Sureshji Khursale, President, Yogeshwari Education Society, Ambajogai who has been a source of constant inspiration. I am especially thankful to Dr. Prakash Prayag, Principal, Yogeshwari Mahavidyalaya, Ambajogai for his encouragement. I owe my special thanks to Dr. P.K. Joshi and Dr. S.P. Chavan of Dnyanopasak Mahavidyalaya, Parbhani; Dr. Mohan S. Kodarkar of Indian Association of Aquatic Biologists, Hyderabad; Dr. Indranil Ghosh of West Bengal University of Animal and Fishery Sciences, Kolkata; Dr. Milind Girkar of College of Fisheries, Udgir; Dr. Meenakshi Jindal of CCS Haryana Agricultural University, Hisar; Dr. Vishwas Shembekar of Rajarshi Shau College, Latur; Shri Sachin Satam of MPEDA, Mumbai; Prof. M.B. Mule of Dr. Babasaheb Ambedkar Marathwada University, Aurangabad; V.B. Mulye of College of Fisheries, Shirgaon, Ratnagiri; Prof. S. Ajmal Khan, Dr. K. Sivakumar and G. Thirumaran of Annamalai University, Parangipettai and Prof. A.K. Patra of Utkal University, Bhubaneshwar.

I wish to thank my wife Surekha for her endurance during the compilation of work of this volume. She has helped me for constantly data feeding, processing and reprocessing on computer. I want to thank my father Shri Balasaheb Sakhare, brother Avinash and sisters Minal and Preeti for their help in many ways. I thank my publisher Shri Anil Mittal of Daya Publishing House, Delhi for taking pains in bringing out the book.

Finally, I will always remain a debtor to all my well-wishers for their blessings, without which this volume would not have come into existence.

Dr. Vishwas Balasaheb Sakhare

Contents

List of Contributors

Anantharaman, P.
Centre of Advanced Study in Marine Biology, Annamalai University, Parangipettai – 608 502, Tamil Nadu

Arul, V.
Department of Biotechnology, Pondicherry University, Kalapet – 605 014, Pondicherry

Arumugam, R.
Centre of Advanced Study in Marine Biology, Annamalai University, Parangipettai – 608 502, Tamil Nadu

Arya, M.
Conservation Biology Unit, School of Studies in Zoology, Jiwaji University, Gwalior

Balaji, K.
Office of the Deputy Director of Fisheries and Fishermen Welfare, 6, Beach Road, Karaikal – 609 602, Puducherry (UT)

Balasubramanian, R.
Centre of Advanced Study in Marine Biology, Annamalai University, Parangipettai – 608 502, Tamil Nadu

Bharti, P.K.
Department of Zoology and Environmental Science, Gurukula Kangri University, Hardwar – 249 404

Bidyalakshmi
Conservation Biology Unit, School of Studies in Zoology, Jiwaji University, Gwalior

Chaudhari, Balaji Sureshrao
Near Sheetal Hotel, Naldurg, Taluka-Tuljapur, District-Osmanabad, Maharashtra

Dande, K.G.
Department of Zoology, Mahatma Basweshwar College, Latur – 413 512, Maharashtra

Devi, Kamala
Zoological Survey of India, Port Blair, Andaman & Nicobar

Garg, R.K.
Conservation Biology Unit, School of Studies in Zoology, Jiwaji University, Gwalior

Garg, S.K.
Department of Zoology and Aquaculture, CCS Haryana Agricultural University, Hisar – 125 004, Haryana

Ghosh, Indranil
Lecturer, Aquaculture Department, WBUAFS, Kolkata, West Bengal
E-mail: ig_1970@yahoo.co.in; wbuafs@gmail.com

Girkar, M.M.
College of Fishery Science, Udgir, Dist. Latur, Maharashtra

Jindal, Meenakshi
Department of Zoology and Aquaculture, CCS Haryana Agricultural University, Hisar – 125 004, Haryana

Kannan, L.
Thiruvalluvar University, Fort Campus, Vellore – 632 004, Tamil Nadu

Karadkhele, S.V.
Department of Zoology, Mahatma Basweshwar College, Latur – 413 512, Maharashtra

Khabade, S.A.
Department of Zoology, D.K.A.S.C. College, Ichalkaranji, Maharashtra

Khan, S. Ajmal
Centre of Advanced Study in Marine Biology, Annamalai University, Parangipettai – 608 502, Tamil Nadu

Kulkarni, A.N.
Department of Zoology and Fishery Science, Science College, Nanded, Maharashtra

Kumar, Prashant
Department of Biotechnology, Pondicherry University, Kalapet – 605 014, Pondicherry

Kumaraguruvasagam, K.P.
Centre of Advanced Study in Marine Biology, Annamalai University, Parangipettai – 608 502, Tamil Nadu

Kushwah, K.
Conservation Biology Unit, School of Studies in Zoology, Jiwaji University, Gwalior

Lokhande, M.V.
Department of Zoology, Mahatma Basweshwar College, Latur – 413 512, Maharashtra

Malik, D.S.
Department of Zoology and Environmental Science, Gurukula Kangri University, Hardwar – 249 404

Mule, M.B.
Department of Environmental Science, Dr. Babasaheb Ambedkar Marathwada University, Aurangabad – 431 004, Maharashtra

Mulye, V.B.
College of Fisheries, Shirgaon, Ratnagiri – 415 629, Maharashtra

Murugan, M.
Centre of Advanced Study in Marine Biology, Annamalai University, Parangipettai – 608 502, Tamil Nadu

Muwal, Manju
Department of Zoology and Aquaculture, CCS Haryana Agricultural University, Hisar – 125 004, Haryana

Patil, P.V.
P.G. Department of Zoology, K.S.K. College, Beed, Maharashtra

Patra, A.K.
Fisheries and Aquaculture Unit, Department of Zoology, Utkal University, Vani Vihar, Bhubaneswar – 751 004, Orissa

Raja, P.
Centre of Advanced Study in Marine Biology, Annamalai University, Parangipettai – 608 502, Tamil Nadu

Raja, S.
Centre of Advanced Study in Marine Biology, Annamalai University, Parangipettai – 608 502, Tamil Nadu

Rao, R.J.
Conservation Biology Unit, School of Studies in Zoology, Jiwaji University, Gwalior
E-mail: soszool@rediffmail.com

Rathod, D.S.
Department of Zoology and Fishery Science, Rajarshi Shau College, Latur – 413 512, Maharashtra

Sadawarte, R.K.
College of Fisheries, Shirgaon, Ratnagiri – 415 629, Maharashtra

Sahu, Maloy Kumar
Centre of Advanced Study in Marine Biology, Annamalai University, Parangipettai – 608 502, Tamil Nadu

Sapkale, P.H.
College of Fisheries, Shirgaon, Ratnagiri – 415 629, Maharashtra

Satam, Sachin
MPEDA, Mumbai

Shembekar, V.S.
Department of Zoology and Fishery Science, Rajarshi Shau College, Latur – 413 512, Maharashtra

Singh, H.
Conservation Biology Unit, School of Studies in Zoology, Jiwaji University, Gwalior

Sivakumar, K.
Centre of Advanced Study in Marine Biology, Annamalai University, Parangipettai – 608 502, Tamil Nadu

Sudhakar, S.
Centre of Advanced Study in Marine Biology, Annamalai University, Parangipettai – 608 502, Tamil Nadu

Sureshrao, Chaudhari Balaji
Near Sheetal Hotel, Naldurg, Tq. Tuljapur, Dist. Osmanabad, Maharashtra State

Taigore, S.
Conservation Biology Unit, School of Studies in Zoology, Jiwaji University, Gwalior

Thangaradjou, T.
Centre of Advanced Study in Marine Biology, Annamalai University, Parangipettai – 608 502, Tamil Nadu

Thirumaran, G.
Centre of Advanced Study in Marine Biology, Annamalai University, Parangipettai – 608 502, Tamil Nadu

Todkari, S.S.
College of Fishery Science, Udgir, Dist. Latur, Maharashtra

Yadav, Rashmi
Department of Zoology and Environmental Science, Gurukula Kangri University, Hardwar – 249 404

Yadava, N.K.
Department of Zoology and Aquaculture, CCS Haryana Agricultural University, Hisar – 125 004, Haryana

Chapter 1

Status of Biodiversity in Inland Wetlands of Gwalior-Chambal Region in Madhya Pradesh

☆ *R.J. Rao, R.K. Garg, S. Taigore, M. Arya, H. Singh, Bidyalakshmi and K. Kushwah*

ABSTRACT

The wetlands are highly productive areas with rich biodiversity, they serve as a spawning and nursery ground for fishes, reptiles, birds, mammals etc. and hence can be used as an excellent area for conservation of rare and endangered species. These wetlands can only be maintained if the ecological processes of wetlands are allowed to continue functioning. Unfortunately, and in spite of important progress made in recent decades, wetlands continue to be among the world's most threatened ecosystems, owing mainly to ongoing drainage, conversion, pollution, and over-exploitation of their resources. To study and evolve remedial measures to the extent possible for conservation of wetlands, various organisations have been conducting a variety of research and planning, implementation and monitoring activities. Remote sensing technology is an important tool in this assessment because of its ability to provide synoptic view of the earth which would not be possible from the ground without exhaustive field surveys. Madhya Pradesh is very rich in wetlands. Majority of the wetland area in Gwalior-Chambal region in North Madhya Pradesh is under man-made category. The water reservoirs and village ponds apart from meeting the requirement of water for agriculture, industry and drinking water also function as aquatic eco-systems. Surveys have been carried out to identify important wetlands in this region to assess the biotic potential of the wetlands and for recommendations for conservation management of priority wetlands.

Introduction

The Indian wetland system has been very broadly classified into two major categories. They are naturally occurring waterlogged areas including flood plains, rivers, lakes etc. and man-made wetlands, including a large number of ponds, small lakes etc., which may be water filled for a variable duration of time. During the past decades wetlands have received increasingly greater attention from the viewpoint of their ecology as well as conservation. The wetlands are now considered to be distinct ecosystems with specific ecological characteristics, functions and values.

Water resources support rich biodiversity. India is drained by numerous rivers, which are fairly well spread. There are many wetlands available in different parts of the country. The freshwater in these rivers and wetlands is amongst the most critical factors limiting urban growth and agriculture development. At present time, wetlands in different areas are used for many purposes. The wetlands are highly productive areas with rich biodiversity, they serve as a spawning and nursery ground for fishes, birds etc. and hence can be used as an excellent area for conservation of rare and endangered species.

The state of Madhya Pradesh straddles the central portion of India, covering an area of around 308,000 sq. km. and falls in two biogeographical zones, semi-arid and the Deccan peninsular. The Madhya Pradesh has 9 national park and 25 wildlife sanctuaries and also has India's largest riverine

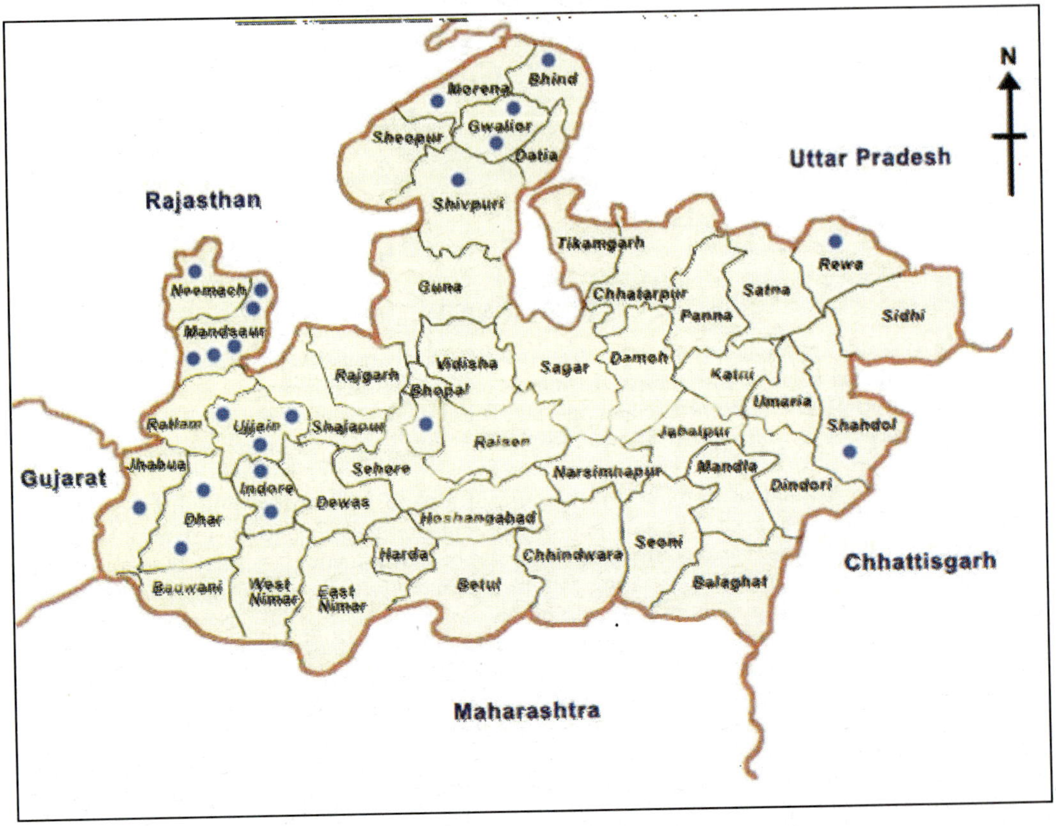

Figure 1.1: Map of Madhya Pradesh, India Showing Districts Surveyed in Northern Part of the State to Locate Important Wetlands

habitats. There are many freshwater bodies in the form of lakes and man-made reservoirs. The reservoirs occupy maximum area (2, 66,863 ha.) followed by tanks, lakes and ash/cooling pond. Valecha, *et al.* (1987) on Bhopal lake, Unni and Patel (1988) on Waghyanala reservoir and Arjariya (2003) conducted studies on physico-chemical profile and plankton in wetlands. Hydro biological studies of Natnagra pound, district Dhar, were carried out by Dhakad and Chaudhary (2005). EPCO (1996) prepared a report on wetlands and urban water bodies in Madhya Pradesh. Rao (2001) compiled all available information on wetlands under UNEP sponsored wetland projects. Limnological studies on the water bodies in and around Gwalior region have also attracted the attention of various workers. The physico-chemical Characteristics of Tekanpur tank (Verma, 1969), Kailasagar (Dagaonkar and Saksena, 1992) have been studied. However, habitat utilization of aquatic animals in these wetlands has not been studied in detail. The biodiversity of certain inland wetlands in Gwalior-Chambal Region is presented in this paper.

Study Area

Surveys have been carried out the in different districts of North Madhya Pradesh. Detailed field visits have been made to several wetlands in Gwalior and Chambal region.

Methodology

There are various approaches to collect information for sustainable use of the wetland resources. The wetlands having high biodiversity values in North Madhya Pradesh have been selected for the biodiversity studies. Regular interaction between the staff in-charge and the research staff on the methodology, collection of required data etc. was maintained. The Survey of India topo maps (1:50,000) were used to prepare the locality map of the wetlands. Biomaps of wetlands were made using Global Position System (GPS – make Garmin 76). Geographical coordinates, elevation, area, water depth at different points, water quality and biotic characteristics at different locations were measured with the help of GPS and other limnological equipments.

Fish samples were collected from the fishermen during early hours of the day from the fish-landing sites at different wetlands. In addition fish was collected by using different nets with the help of local fishermen during different hours of the day in different seasons. The fish was brought to the laboratory for identification purpose. Information on fish production from different wetlands was obtained from the Fisheries Department. Interviews with the fish contractors and from local fishermen were made to collect information on fish production.

Different species of aquatic animals and migratory birds were identified through direct sightings, collection of live specimens from the wetlands. All species were identified systematically.

Results

In the Gwalior-Chambal Region comprising 6 six districts a total of 15 important wetlands have been identified. These wetlands are highly used for various purposes like drinking, irrigation, fisheries development. They are ideal habitat for migratory birds, several aquatic animals including mugger crocodile, freshwater turtles (*Lissemys punctata, Pangshura tecta*), amphibians, fishes, molluscs, benthic fauna and planktons. Among the identified wetlands in the region Chandpata lake, Dihaila lake and Tighra Reservoir are located inside protected areas.

The list of wetlands surveyed is given in Table 1.1. Details of climatic conditions, vegetation, fish resources, avifaunal resources, and socio-economic status of several wetlands located in different districts of Madhya Pradesh are shown in Table 1.2-1.5.

Table 1.1: Human Usage of Various Water Bodies in the North Madhya Pradesh

Sl.No.	Wetland	District	Human Use
1.	Tighra Reservoir	Gwalior	Pisciculture, Agriculture, Drinking
2.	Ramoua Reservoir	Gwalior	Irrigation, Drinking, Pisciculture
3.	Sirsa Reservoir	Gwalior	Irrigation, Pisciculture
4.	Harsi Reservoir	Gwalior	Irrigation, Pisciculture
5.	Kaketo	Gwalior	Irrigation, Pisciculture
6.	Pehsari Reservoir	Gwalior	Irrigation, Pisciculture
7.	Tekankpur Lake	Gwalior	Irrigation, Fishing
8.	Pagara Reservoir	Morena	Irrigation, Pisciculture
9.	Chandpata Lake	Shivpuri	Drinking, Tourism
10.	Dihaila Lake	Shivpuri	Irrigation
11.	Piluwa Reservoir	Morena	Irrigation, Pisciculture
12.	Kotwal Reservoir	Morena	Irrigation, Pisciculture
13.	Gohad Talab	Bhind	Irrigation, Pisciculture
14.	Tekanpur lake	Gwalior	Drinking, Irrigation, Pisciculture
15.	Ramsagar lake	Datia	Drinking, Irrigation, Pisciculture

Table 1. 2: Geographical Location of Several Wetlands in North Madhya Pradesh

Sl.No.	Name of the Wetland	District	Latitude	Longitude	Altitude (m)
1	Ramsagar	Datia	N25°40'4895"	E78°23'8887"	228.9
2	Tighra (Saank R.)	Gwalior	N26°12'9354"	E78°01'0126"	221.0
3	Sirsa	Gwalior	N26°02'7069"	E77°53'2550"	309.7
4	Kaketo (Parbati R)	Gwalior	N25°53'8774"	E77°52'0572"	345.2
5	Pesari	Gwalior	N25°57'9339"	E77°46'0864"	324.1
6	Harsi (Parbati R)	Gwalior	N22°40'3045"	E77°54'0028"	327.6
7	Kotwal (Aasun R)	Morena	N26°29'5475"	E78°10'7549"	228.5
8	Piluwa (Sank R)	Morena	N26°28'5447"	E78°12'7549"	268.5
9	Pagara (Aasun R)	Morena	N26°16'5687"	E77°51'7239"	258.5
10	Chandpata	Shivpuri	N26°00'9153"	E77°52'9235"	329.1
11	Dihaila	Shivpuri	N25°27'2903"	E78°04'4879"	316.3

Species Diversity

Madhya Pradesh wetlands shelter many species of fauna and flora, of which the most widely explored, scientifically studied and appreciated are the aquatic reptiles and birds. Many wetlands are renowned because of their birdlife.

The aquatic biodiversity in the Gwalior-Chambal region is very rich. About 105 fish species of which 39 species are of commercial importance have been reported in North M.P. Important species of major carps include *Catla, Labeo, Cyprinus,* and *Cirhinus*. The Catfishes include *Mystus, Wallago,*

Heteropneustus, Pungasius, Notopterus etc. There is one species of crocodile (*Crocodylus palustris*), two species of freshwater turtles {*Aspideretes gangetics,* and *Lissemys punctata*) and more than 120 species of wetland birds present in different wetlands and rivers in the region (Table 1.2).

Table 1.3: List of Fish Species Recorded in Ramsagar Reservoir

ORDER / *Family*	*SUBFAMILY* / *Genus* / *Species*
OSTEOGLOSSIFORMES	
Notopteridae	
	Notopterus Lacepede
	1. *N. notopterus* (Palas)
CYPRINIFORMES	
Cyprinidae	DANIONINAE
	Salmostoma Swainson
	2. *S. bacaila* (Ham.-Buch.)
	3. *S. clupeoides* (Bloch)
	Barilius Hamilton-Buchanan
	4. *B. barila* (Ham.-Buch.)
	5. *B. bola* (Ham.-Buch.)
	6. *B. bendelisis* (Ham.-Buch.)
	Rasbora Bleeker
	7. *R. daniconius* (Ham.-Buch.)
	Danio Hamilton-Buchanan
	8. *D. devario* (Ham.-Buch.)
	CYPRININAE
	Tor Gray
	9. *T. tor* (Ham.-Buch.)
	Puntius Hamilton-Buchanan
	10. *P. conchonius* (Ham.-Buch.)
	11. *P. sarana* (Ham.-Buch.)
	12. *P. sophore* (Ham.-Buch.)
	13. *P. ticto* (Ham.-Buch.)
	Osteobrama Heckel
	14. *O. cotio cotio* (Ham.-Buch.)
	Catla Valenciennes
	15. *C. catla* (Ham.-Buch.)
	Cirrhinus Oken
	16. *C. mrigala* (Ham.-Buch.)
	17. *C. reba* (Ham.-Buch.)

Contd...

Table 1.3–Contd...

ORDER / Family	SUBFAMILY / Genus / Species
	Labeo Cuvier
	18. *L. bata* (Ham.-Buch.)
	19. *L. calbasu* (Ham.-Buch.)
	20. *L. gonius* (Ham.-Buch.)
	21. *L. rohita* (Ham.-Buch.)
	GARRINAE
	Garra Hamilton-Buchanan
	22. *G. gotyla.gotyla* (Gray)
Balitoridae	NEMACHEILINAE
	Acanthocobitis Peters
	23. *A. botia* (Ham.-Buch.)
SILURIFORMES	
Bagriidae	BAGRINAE
	Mystus Scopoli
	24. *M. bleekeri* (Day)
	25. *M. tengara* (Ham.-Buch.)
	Aorichthys Wu
	26. *A. seenghala* (Sykes)
	27. *A. aor* (Ham.-Buch.)
Siluridae	
	Ompok Lacepede
	28. *O. bimaculatus* (Bloch)
	Wallgo Bleeker
	29. *W. attu* (Bloch and Schneider)
Schilbeidae	SCHILBEINAE
	Eutropiichthys Bleeker
	30. *E. vacha* (Ham.-Buch.)
	Silonia Swainson
	31. *S. silondia* (Ham.-Buch.)
Sisoridae	
	Bagarius Bleeker
	32. *B. bagarius* (Ham.-Buch.)
Ciariidae	
	Clarias Scopoli
	33. *C. batrachus* (Linnaeus)

Contd...

Table 1.3–Contd...

ORDER / *Family*	*SUBFAMILY* / *Genus* / *Species*
Heteropneustidae	
	Heteropneustes Muller
	34. *H. fossilis* (Bloch)
BELONIFORMES	
Belonidae	
	Xenentodon Regan
	35. *X. cancila* (Ham.-Buch.)
SYNBRANCHIFORMES	
Mastacembelidae	MASTACEMBELINAE
	Mastacembelus Scopoli
	36. *M. armatus* (Lacepede)
PERCIFORMES	
Chandidae	
	Chanda Hamilton-Buchanan
	37. *C. nama* (Ham.-Buch.)
Nandidae	NANDINAE
	Nandus Valenciennes
	38. *N. nandus* (Ham.-Buch.)
Gobiidae	GOBIINAE
	Glossogobius Gill
	39. *G. giuris* (Ham.-Buch.)
Channidae	
	Channa (Ophiocephalus) Scopoli
	40. *C. marulius* (Ham.-Buch.)
	41. *C. striata* (Bloch)
	42. *C. punctatus* (Bloch)

In India many bird species are wetland-dependent. Although many of these birds are known, much about their habitats remains uninvestigated. There are two categories of water-birds: wetland specialists and generalists. Specialists are those that nest, feed and roost in wetlands. Wetland specialists are wholly dependent on aquatic habitats, and cannot survive without them. Examples are ducks, gulls, herons, waders, etc. Generalists are those birds that are frequently found in wetlands, but are sometimes seen in other habitats as well, such as ibises, herons, some weavers, warblers, plovers etc.

Table 1. 4: Fish Species Diversity in Ramsagar Reservoir

Sl.No.	Families	Genus	Species	Per cent of Contribution of Families
1.	Notopteridae	01	01	2.38
2.	Cyprinidae	11	21	50.00
3.	Balitoridae	01	01	2.38
4.	Bagriidae	02	04	9.52
5.	Siluridae	02	02	4.76
6.	Schilbeidae	02	02	4.76
7.	Sisoridae	01	01	2.38
8.	Clariidae	01	01	2.38
9.	Heteropneustidae	01	01	2.38
10.	Belonidae	01	01	2.38
11.	Mastacembelidae	01	01	2.38
12.	Chandidae	01	01	2.38
13.	Nandidae	01	01	2.38
14.	Gobiidae	01	01	2.38
15.	Channidae	01	03	7.14
	Total	28	42	

Table1.5: List of Wetland Birds Found in the Wetlands of Gwalior/Chambal Region

Sl.No.	Birds	1	2	3	4	5	6	7	8	9	10	11	12	13
1.	Asian Open bill		✓								✓			✓
2.	Bar headed Goose	✓		✓		✓	✓			✓	✓		✓	✓
3.	Black Ibis	✓									✓			✓
4.	Black-winged Stilt	✓	✓	✓	✓	✓	✓	✓	✓	✓	✓	✓	✓	✓
5.	Cattle Egret	✓	✓	✓	✓	✓	✓	✓	✓	✓	✓	✓	✓	✓
6.	Comb Duck	✓					✓				✓			
7.	Common Crane							✓			✓			
8.	Common Pochard	✓		✓	✓						✓			
9.	Common Sandpiper	✓	✓	✓	✓	✓	✓	✓	✓	✓	✓	✓	✓	✓
10.	Common Snipe	✓			✓	✓					✓	✓		
11.	Common Teal													
12.	Coots	✓	✓				✓	✓		✓	✓			✓
13.	Curlew Sandpiper	✓	✓	✓	✓			✓				✓		
14.	Darter (Snake bird)	✓					✓				✓			✓
15.	Egyptian Vulture	✓			✓	✓	✓		✓	✓	✓	✓		✓
16.	Eurasian Curlew	✓			✓						✓		✓	
17.	Eurasian Thick-knee	✓	✓								✓			
18.	Eurasian Wigeon	✓	✓		✓		✓		✓		✓	✓		✓

Contd...

Table1.5–Contd...

Sl.No.	*Birds*	*1*	*2*	*3*	*4*	*5*	*6*	*7*	*8*	*9*	*10*	*11*	*12*	*13*
19.	Gadwall	✓			✓		✓		✓		✓			
20.	Great Cormorant	✓	✓	✓	✓	✓	✓	✓	✓	✓	✓	✓	✓	✓
21.	Great Egret	✓	✓	✓	✓	✓	✓	✓	✓	✓	✓	✓	✓	✓
22.	Great Thick-knee	✓						✓			✓			
23.	Green Sandpiper		✓	✓										
24.	Grey Heron	✓			✓	✓	✓	✓	✓	✓	✓	✓	✓	✓
25.	Grey Wagtail	✓	✓	✓	✓	✓	✓	✓	✓	✓	✓	✓	✓	✓
26.	Indian Cormorant	✓	✓	✓	✓	✓	✓	✓	✓	✓	✓	✓	✓	✓
27.	Indian Pond-Heron	✓	✓	✓	✓	✓	✓	✓	✓	✓	✓	✓	✓	✓
28.	Intermediate Egret		✓	✓										
29.	Large Cormorant	✓			✓	✓	✓	✓	✓	✓	✓	✓	✓	✓
30.	Lesser Flamingo	✓	✓	✓	✓	✓	✓	✓	✓	✓	✓	✓	✓	✓
31.	Lesser whistling Teals	✓									✓			
32.	Little Bittern		✓	✓										
33.	Little Cormorant	✓	✓	✓	✓	✓	✓	✓	✓	✓	✓	✓	✓	✓
34.	Little egrets	✓			✓	✓	✓	✓	✓	✓	✓	✓	✓	✓
35.	Little Stint	✓	✓	✓	✓						✓			
36.	Little Swift	✓		✓				✓			✓			
37.	Little Tern	✓	✓	✓	✓	✓	✓	✓	✓	✓	✓	✓	✓	✓
38.	Mallard	✓			✓	✓					✓			
39.	Nakta	✓	✓		✓			✓			✓		✓	
40.	Northern Pintail	✓			✓		✓				✓		✓	✓
41.	Northern Shoveler	✓	✓				✓				✓			
42.	Open billed Stork	✓									✓			✓
43.	Osprey	✓										✓		
44.	Painted Stork	✓			✓		✓			✓	✓		✓	✓
45.	Pintail	✓			✓	✓	✓	✓			✓		✓	✓
46.	Purple Heron		✓											
47.	Red crested Pochard	✓			✓	✓	✓	✓			✓		✓	✓
48.	Red-wattled Lapwing	✓	✓	✓	✓	✓	✓	✓	✓	✓	✓	✓	✓	✓
49.	River Lapwing	✓	✓	✓	✓	✓	✓	✓	✓	✓	✓	✓	✓	✓
50.	Ruddy Shelduck	✓	✓	✓	✓	✓	✓	✓	✓	✓	✓	✓	✓	✓
51.	Sarus Crane	✓					✓				✓		✓	✓
52.	Scavanger Vulture	✓			✓		✓	✓						
53.	Spoon bill	✓									✓		✓	✓
54.	Spot bill Duck	✓					✓				✓			
55.	White breasted kingfisher	✓			✓	✓	✓	✓	✓	✓	✓	✓	✓	✓
56.	Woolly-necked Stork		✓			✓						✓		

1: Tighra; 2: Ramoua; 3: Sirsa; 4: Harsi; 5: Kaketo; 6: Pehsari; 7: Takenpur; 8: Pagara; 9: Chandpata; 10: Dihaila; 11: Piluwa; 12: Gohad; 13: Ram Sagar.

Threats

Biodiversity is being lost or endangered in many wetlands in the study area because of habitat degradation and over use of resources. The biodiversity and the climate are intrinsically linked, and likewise, the issues of Biodiversity loss and climate changes are intertwined. The changes in climate affect the wetlands substantially by influencing availability of water in these wetlands. Higher temperatures and less rainfall may lead to drying of wetlands and bring about degradation of water quality. This resulted into Biodiversity loss in these wetlands. The populations of wetland species have been effected greatly by changes in the water cycle. The timing of hydrological events that changed due to climate changes has a major implication on the wetland ecosystem and dependent communities.

Discussion

Although a few freshwater sites were listed in the Project Aqua (Luther and Rzoska, 1971), it was only after; India became a signatory to the Ramsar Conservation in 1981, that efforts were initiated to prepare an inventory of wetlands. An earlier report on the estimates of wetland resources of the country (Biswas, 1976) included all large lakes and reservoirs as well as small ponds and temple tanks as wetlands. One of the simplest classifications is that of the IUCN which has been used in their inventory of Asian wetlands (Scott, 1989). Gopal and Shah (1995) proposed a simple hierarchical classification of wetlands based on their location salinity, physiognomy, duration of flooding and the growth forms of the dominant vegetation. Depending on the species diversity, habitat features of wetlands studied in the Gwalior-Chambal region, the Tighra, Pehsari, Dihalia, Chandpata reservoirs are prioritized as important wetlands and need conservation management of these water bodies.

In the present study it is evident that locals use the water body primarily for drinking, irrigation and also for fishing purpose. Majority of the water bodies have been promoted for the development of fisheries. The fishermen cooperatives have rights on the fishing activities in this reservoir. Killing of migratory birds was also reported. Locals are not aware of the importance of migratory birds and the legislation. Due to mixing of pesticides through water run off from the agriculture fields the water bodies have been infected. Direct causes on site are: (*i*) Overexploitation, (*ii*) Conversion of habitats, (*iii*) destructive land-use practices and (*iv*) Pollution. Some of the greatest threats are diversion of water for different uses, loss of riparian vegetation etc. Therefore, specific actions are needed for conservation of water bodies in the Gwalior-Chambal Region, North Madhya Pradesh, India.

Acknowledgements

We thank our Vice Chancellor; Coordinator, UGC-SAP; and Head, SOS Zoology for permission and encouragement to conduct the present study and UGC for financial assistance.

References

Arjariya, A., 2003. Physico-chemical profile and plankton diversity of Ranital Lake, Chhatarpur, M.P. *Nat. Env. Poll. Tech.*, 2(3): 327–328.

Biswas, B., 1976. India: National report. In: *Proceedings of the International Conference on Conservation of Wetlands 2nd Waterfowl*, (Ed.) M. Smat. Heiligenhafen. West Germany. International Waterfowl Research Bereau. Slimbridge, U.K., p. 108–109.

Dagaonkar, A. and Saksena, D.N., 1992. Physico-chemical and biological characterization of a temple tank, Kailasagar, Gwalior, Madhya Pradesh. *J. Hydrobiol.*, 8(1): 11–19.

Dhakad, N.K. and Chaudhary, P., 2005. Hydrobiological study of Natnagra pond in Dhar district (M.P.) with special reference to water quality impact on potability, irrigation and aquaculture. *Nat. Env. Poll. Tech.*, 4(2): 269–272.

EPCO, 1996. *Third Environmental Status Report*, EPCO, Bhopal, Madhya Pradesh, India.

Gopal, B. and Shah, M., 1995. Inventory and classification of wetlands of India. *Vegetatta* 118: 39–48.

Luther, J. and Rzoska, J., 1971. *Project Aqua: A Source Book of Inland Waters Proposed for Conservation*, (Eds.) J. Luther and J. Rzoska. IBP Handbook No. 21, Blackwells, Oxford.

Rao, R.J., 2001. Workshop on inland wetlands of Madhya Pradesh and Chhattishgarh. *All India Coordinated Project Inland Wetland of India*, Briefing Book.

Scott, D.A. Ed., 1989. *A Directory of Asian Wetlands*, IUCN, Gland, Switzerland, p. 1181.

Unni, K.S. and Patel, M.K., 1988. Limnology of an eutrophic deep discharge reservoir in central India. *Proc. Nat. Symp. Past, Present and Future of Bhopal Lakes*, p. 137–140.

Valecha, V., Trivedi, R. and Bhatnagar, G.P., 1987. Biological assessment of trophic status of lower lake of Bhopal. *Poll. Res.*, 6: 91–93.

Verma, M.N., 1969. Hydrobiological study of a tropical impoundment Tekanpur reservoir, Gwalior, India with special reference to the breeding of Indian Carps. *Hydrobiologia*, 34: 358–368.

Chapter 2

Marine Ornamental Fishes in the Little Andaman Island

☆ *M. Murugan, Maloy Kumar Sahu, M. Srinivasan, Kamala Devi, S. Ajmal Khan and L. Kannan*

Introduction

Fishes form rich sources of food and also provide several by-products to us. So they are extremely important and economically valuable resources. The seas around the Andaman and Nicobar group of Islands are well known for their rich fishery resources (Talwar, 1990; Mehta and Kamala Devi, 1990; Rajan *et al.*, 1992, 1993; Rao *et al.*, 1992, 2000; Dhandapani and Mishra, 1993, 1998; Rao and Kamala Devi, 1996; Kamala Devi and Rao, 1997; Gosh, 2001). The ornamental fishes occurring in the fishes of the Little Andaman Island, are capable of meeting the commercial interests development in this area. Occurrence of most interesting and fascinating fishes in the Little Andaman waters is due to the diversity of marine habitats such as mangroves, creeks, muddy shores, coral reefs etc. However our knowledge on ornamental fish resources of the Little Andaman Island is very limited. Hence, an attempt has been made to study the ornamental fish resources of the Little Andaman Island.

Materials and Methods

Collections were made around the Island using various fishing gears like shore seine, gillnet and cast net in the main fishing places like Hut Bay, Naval Area, Harbindar Bay, Chandranallah Coast, Dugong Creak, Butler Bay, Nethaji Nagar Coast and Nanjappa Nagar Coast. In addition, specimens were also procured from the fish market at Hut Bay and landing centers in Nethaji Nagar and Nanjappa Nagar. The specimens were preserved in 10 per cent formaldehyde and analyzed intensively in order to place the species correctly in their taxonomic position. Specimens were identified using the standard works of Allen and Steene (1987), Masuda *et al.* (1984), Talwar and Kacker (1984), Abu Khair Mohammad *et al.* (1996); Rao (2003); Rajan (2003).

Results and Discussion

During the present investigation five different species of ornamental fishes were recorded from the Island. Their systematic position, distribution, characters, colour, biology and fishery are given below.

Butterfly Fish

Systematic Position

Order: Perciformes

Family: Chaetodontidae

Genus: Chaetodon

Species: Chaetodon vagabundus *(Linnaeus, 1758)*

Distinguishing Characters

The body of this butterfly fish is compressed and somewhat rounded and very deep with its depth being 1.5 to 1.6 times its standard length. Ctenoid scales cover the body. There are 29 to 35 scales along the midline. The majority of the trunk scales are relatively large and rhomboid. The fins include a single continuous dorsal fin with 12 to 13 spines and 23 to 26 rays, an anal fin with 3 spines and 19 to 22 rays and a rounded caudal fin. The spinous dorsal base is longer than the soft dorsal base. The mouth is terminal and protruding and bears five bristle-like teeth in both jaws. There are 4 to 6 gill rakers on the upper limb and 11 to 15 on the lower limb of the first gill arch respectively.

Colour

The overall body colour of this fish is pearly white becoming yellowish posteriorly. The body in front and at the top, has about 8 streaks or lines directed obliquely upwards and backwards. This set of streaks is met at an angle by another set of about 12 lines running downwards and backwards. A broad black bar running through the eye is prominent. There are further two black bars, one a blackish crescentic bar, runs along the bases of the dorsal, caudal and anal fins, and the other which has a narrow white edge on the anal fin. The caudal fin is yellow.

Distribution

The species occurs throughout tropical Indo-West Pacific region.

Biology and Fishery

Inhabiting shallow coral rocky reefs down to 30 meters depth and is usually encountered in pairs. The diet of this species includes small crabs marine worms and other small invertebrates. The vagabond butterfly fish attains 23 centimeters in total length.

Silver Moony

Systematic Position

Order: Perciformes

Family: Monodactylidae

Genus: *Monodactylus*

Species: *Monodactylus argenteus* (Linnaeus, 1758)

Distinguishing Characters

The body of this diamond-shaped fish is strongly composed and very deep, its depth being 1.3 to 1.5 times of its standard length. The mouth, which is small and terminal bears very fine villiform teeth in each jaw. The dorsal and anal fins are symmetrical with dorsal fin comprising of 7 or 8 spines and 26 to 30 soft rays and the anal fin with 3 spines is small and inconspicuous. The pectoral fins are short and rounded and have 16 rays each. The large caudal fin is truncate to slightly emarginated. The body is covered with small deciduous ctenoid scales, which are arranged in 52 to 60 series along the lateral line. Scales also extend out to the fins. There are 25 to 28 gill rakers on the first gill arch with 6 to 8 rakers occurring on the upper limb and 19 to 22 on the lower limb of the gill arch.

Colour

As the name implies, the silver moony is silvery to silvery gray overall with its fins having yellowish tips. The dorsal and anal fin lobes are dusky to black. Young specimens have two dusky bars across the head which fade with age.

Distribution

This fish, which is widespread in the tropical Indo–Pacific is widely distributed in Omani waters occurring along the entire coastline except in the Arabian Gulf.

Biology and Fishery

The silver moony is gregarious and is commonly found in small but dense shoals in estuaries, wharfs and shallow reefs. It can tolerate wide fluctuations in salinity and may occur in freshwater also. It mainly feeds in the water column and its diet consists of planktonic organisms and detritus. It is believed to attain sexual maturity at about 15 to 18 cms in length and to lay few numbers of large eggs. It uses estuaries as a nursery area. The silver moony is a small fish, growing upto 25 centimeters in total length.

Silver Moony

Systematic Position

Order: Perciformes

Family: Caesionidae

Genus: *Caesio caerulaurea* (Lacepede, 1801)

Species: *Caesio caerulaurea* (Lacepede, 1801)

Distinguishing Characters

Body elongate, slender and slightly compressed. Dorsal and ventral profile equally convex. Anterior region of head and lower region of preoperculum and operculum scaleless. Interorbital space slightly concave, width about 1.3 times eye diameter. Eye moderately large, adipose eyelid not well developed, covering only orbital margin of eye. Eye diameter greater than snout length. Snout slightly blunt. Small terminal mouth within thin lips. A series of minute conical teeth in both jaws. No teeth on palatine and vomer. Maxilla reaching below anterior margin of pupil. Head and body covered with small ctenoid scales. On head supratemporal bands of scales interrupted at dorsal mid-line by a scales zone, 'V' shaped scaleless zone anteriorly at mid-line intruding between supratemporal band of scales. Lateral line running horizontally. Dorsal and anal fins almost completely scaled. Pelvic fin with axillary scales. Dorsal fin with minute first spine. Caudal fin deeply forked.

Colour

Upper half of head and body bright metallic blue and lower half of body pinkish. Golden horizontal band running from head to caudal fin base, just above lateral line. A black blotch at upper base of pectoral fin. Broad black band on each lobe of caudal fin.

Distribution

Found in Indo-Pacific region including Malaysia.

Biology and Fishery

Very common in coral reef areas and usually seen in large schools. Prefers swimming at the surface of water or under floating objects and boats. Feeds mainly on zooplankton, crustaceans and small fishes; common size about 20 cm, maximum 35 cm. Usually caught using purse seines and traps.

Convict Surgeon Fish

Systematic Position

Order: Perciformes

Family: Acanthuridae

Genus: *Acanthurus*

Species: *Acanthurus lineatus* (Linnaeus, 1758)

Distinguishing Characters

Body deep and compressed. Body depth 2.2 times in standard length. Mouth small with spatulate teeth, close-set, with denticulate edges. Caudal peduncle with a lancet-like spine folded into deep grooves on each side. Caudal spine long and venomous. Stomach thin-walled. Caudal fin deeply lunate, with filamentous upper and lower rays.

Colour

Body and head with alternating blue and yellow black-edged stripes except for lower ¼, which is bluish-white. Dorsal fin with pale blue and yellowish stripe. Anal fin grey, yellow basally with light blue edge. Caudal fin blackish with a grey crescent centro-posteriorly, front with bluish-white edge and black margins. Pectoral fins pale. Pelvic fins orange-yellow with white lateral margin and black sub marginally.

Distribution

Found in the Indo-Pacific, the South China Sea and western Indian Ocean southward to Mozambique.

Biology and Fishery

Inhabits inshore waters feeding on benthic algae on coral reefs or rocky areas. Known to be an aggressive territorial fish. Common size caught 25 cm, maximum length 30cm. Caught by cast nets, gill nets, spears and traps.

Blue Striped Surgeon Fish

Systematic Position

Family: Acanthuridae

Genus: *Acanthurus*

Species: *Acanthurus triostegus*

Distinguishing Characters

Body high and compressed. Body depth 1.8 times in standard length. Dorsal profile of head before eye concave, above eye, convex. Mouth small, teeth spatulate, close-set, with denticulated edges. Dorsal fin continuous, intermembrane without notch. Caudal peduncle with a lancet-like spine which folds into a deep groove. Caudal fin truncate to slightly emarginated.

Colour

White-greenish dorsally and white ventrally. 5 dark vertical bars, 1st from nape passes through eye downward, 2nd from 1st spine of dorsal fin to base of pectoral fin, 3rd –5th bars below dorsal fin and 1 on caudal peduncle.

Distribution

Found in east Africa to the east coast of Mexico, Japan, Indo-Pacific and the South China Sea.

Biology and Fishery

Inhabits inshore waters and feeds on algae. Dwells either singly or in large groups. Spawning occurs at dusk, from 12 days before to 2 days after full moon. Eggs are pelagic and spherical with a single oil droplet. Average diameter of eggs is about 0.68 mm. Eggs hatch after 26 hours. Juveniles live in very shallow waters of tide pools or reef flats and grow at about 12 mm per month. Common length caught 17 cm; maximum length 24 cm.

Rajan *et al.* (1992) studied the butterfly fishes occurring in Andaman and Nicobar Islands. Rao and Kamala Devi. (1996) and Rajaram *et al.* (2004) recorded 5 species from Great Nicobar Islands. In the present study 6 important ornamental fishes have been recorded, *viz. Chaetodon vagabundus, Monodactylus argenteus, Caesio caerulaurea, Acanthurus lineatus, A. triostegus* and *Epinephalus malabaricus* which can be used in ornamental fish trade.

The various suggestions (Sundararaj, 2005) made for preserving the resources of marine ornamental fishes includes:

1. Dredging and blasting of coral growing areas or their adjacent areas for navigational and harbour activities should be avoided.
2. Collection of coral, removal of surface soil, cutting of natural vegetation from the shore and polluting the shore by oil and other substances should be prevented.
3. Fishing methods, which damage the ecosystem should not be practiced in coral growing areas.
4. Collection of marine ornamental fishes for aquarium fish trade should follow eco-friendly methods.
5. Exploitation of ornamental fishes should be within limits.
6. Research on captive breeding and rearing of marine ornamental fishes should be promoted.
7. Supply of hatchery produced marine ornamental fishes to the aquarium fish trade, should become a reality.
8. Support facilities must be provided and specific scientific groups including private sector should be brought together to achieve success in the production and marketing of marine ornamental fishes both within the country and abroad.
9. For coastal area development, rewarding programme must be taken up with in greater funding support on ornamental fish breeding and seed production, developing facilities along the coastal areas and

10. A center for marine ornamental fish must be established in the country with representing branches.

There are enormous opportunities to develop ornamental fish production and trade in India. Ornamental fish collection, breeding, seed production and marketing can create huge employment opportunities for about 0.5 million people annually in this sector. Institutes of technology development joining with the agencies of human resource development should come out with a massive developmental programme at national level and open new avenues for alternative employments along the coast.

Acknowledgments

Authors thank Prof. T. Balasubramanian, Director, Centre of Advanced Study in Marine Biology for the encouragement and the authorities of Annamalai University for facilities. They also thank the Ministry of Environment and Forests, Government of India for the financial support to carry out the work in the Little Andaman Island.

References

Dhandapani, P. and Mishra, S.S., 1993. New records of marine fishes from Great Nicobar. *J. Andaman Sci. Assoc.*, (9): 58–62.

Gerald Allen, R. and Steene, Roger C., 1987. *Reef Fishes of the Indian Ocean*, USA.

Ghosh, S.K., 2001. Andaman and Nicobar Islands: Untapped fishery resources. *Bay of Bengal News*, p. 18–21.

Kamaladevi and Rao, D.V., 1997. New records of reef fishes from Andaman waters. *J. Andaman Sci. Assoc.*, 13: 104–106.

Masuda, H., Amaoka, K., Araga, C., Uyeno, T. and Yoshino, T., 1984. *The Fishes of the Japanese Archipelago.* Tokai University Press, Japan, p. 437.

Mehta, H.S., Rajan, P.T. and Kamaladevi, 1990. Five new records of fishes from Bay Islands. *J. Andaman Sci. Assoc.*, 6: 193–195.

Mehta, H.S. and Kamaladevi, 1990. Four new records of Gobioid fishes from Andaman and Nicobar Islands. *J. Andaman Sci. Assoc.*, 6: 66–68.

Mohammed Mohsin, A.K. and Ambak, Mohammed Azmi, 1996. *Marine Fishes and Fisheries of Malaysia and Neighboring Countries*, p. 1–687.

Rajan, P.T., 2003. *A Field Guide to Marine Food Fishes of Andaman and Nicobar Islands.* Zoological Survey of India, Kolkata, p. 1–260.

Rajan, P.T., Rao, D.V., Kamaladevi and Dey, S., 1993. New records of rare fishes from Andaman Islands. *J. Andaman Sci. Assoc.*, 9: 103–106.

Rajan, P.T., Rao, D.V. and Kamaladevi, 1992. New records of butterfly fishes from Andaman and Nicobar Islands. *J. Andaman Sci. Assoc.*, 8: 172–174.

Rajaram, R., Srinivasan, M., Ajmalkhan, S. and Kannan, L., 2004. Ichthyofaunal diversity of Great Nicobar Island, Bay of Bengal. *J. Indian Fish. Assoc.*, 31: 13–26.

Rao, D.V. and Kamaladevi, 1996. Notes on rabbit fishes (Family: Siganidae) of Andaman and Nicobar Islands. *J. Andaman Sci. Assoc.*, 12: 84–88.

Rao, D.V., Kamaladevi and Rajan, P.T., 2000. An Account of ichthyofauna of Andaman and Nicobar Islands, Bay of Bengal. *Records of the Zoological Survey of India*, Occ. Paper No. 178: 1–434.

Rao, D.V., Rajan, P.T. and Kamaladevi, 1992. New records of groupers (Family: Serranidae) and cardinal fishes (Family: Apogonidae) from Andaman and Nicobar Islands. *J. Andaman Sci. Assoc.*, 8: 47–52.

Rao, D.V., 2003. *Guide to Reef Fishes of Andaman and Nicobar Islands*. Zoological Survey of India, Kolkata, p. 1–555.

Sundararaj, V. and Satheesh, J.M., 2005. Tropical marine aquarium resources of marine ornamental fishes. *ENVIS Center*, p. 29–38.

Talwar, P.K. and Kacker, R.K., 1984. In: *Commercial Sea fishes of India*. Handbook No. 3, Zoological Survey of India, 997 pp.

Talwar, P.K., 1990. Fishes of Andaman and Nicobar Islands: A synoptic analysis. *J. Andaman Sci. Ass.*, 6: 71–102.

Chapter 3

Influence of Supplementary Feeds on the Growth and Excretory Metabolite Levels in *Heteropneustes fossilis*

☆ *Meenakshi Jindal*

ABSTRACT

The fingerlings of *Heteropneustes fossilis* were fed on 10 formulated diets (protein levels ranged between 35-45 per cent) over a period of 80 days to study their protein requirements and effect of diets on the excretory levels of ammonical nitrogen (NH_4-N) and ortho-phosphate (O-PO_4) by using 2 protein sources *i.e.* processed soybean; HPS and fishmeal; FM. Growth and Specific growth rate remained low when fed on low protein levels (*i.e.* 35 per cent and 38 per cent). Highest growth performance was observed in groups fed on 40.25 per cent dietary protein irrespective of the protein source, but better growth was observed in the fish group fed on 40.25 per cent HPS based diets as compared to 40.25 per cent FM based diets. An increase in dietary protein level (beyond 40.25 per cent) showed repressed growth. The values of DO and excretion of NH_4-N and O-PO_4 remained low where fish were fed on 40.25 per cent dietary protein irrespective of the protein source. But these values remained high for groups fed on FM based diets. The results clearly showed that out of two protein sources used in these studies, soybean gaves the most promising results. Therefore, the use of soybean (at 40.25 per cent protein level) is recommended as an eco-friendly and cost effective dietary protein source in the diets of *H. fossilis*.

Introduction

Protein is typically the most costly nutrient in a formulated feed. Feed costs are usually the major operational expense in most aquacultural operations, typically ranging from 30-50 per cent of the variable operating costs. Protein in the majority of formulated fish diets (worldwide) depends greatly on fishmeal (FM), which is more costly than high quality plant-based protein sources, such as soybean, canola, cotton seed meal, moong etc. Due to world wide dominance of soybean and its appreciation as quality protein, several workers have attempted to replace FM with soybean meal (SBM) in diets formulated for several fish species (Jindal, 2008; Jindal and Garg, 2005; Jindal *et al.*, 2007a,b; Robinson and Menghe, 2007).

The earlier studies of Singh (2001) have revealed that diets consisting of only rice bran and oil cake commonly used by farmers neither contain essential nutrients in sufficient quantity nor optimum protein levels. These studies have also revealed that growth and survival appeared to a function of dietary protein level, indicating high dietary protein requirements of fishes under field conditions. Since in these studies, highest dietary protein levels were only 35 per cent, therefore, to determine the optimum protein requirements of *H. fossilis* fingerlings, diets containing 35–45 per cent crude protein were formulated and fed to the fingerlings under laboratory conditions. In the present studies, therefore, efforts were also made to compare the efficacy of FM versus soybean containing diets at 40 per cent dietary protein level.

Nitrogen in waste water from aquaculture effluents is often considered a pollutant. In freshwater systems, nitrogen is sometimes a limiting nutrient, so adding it stimulates plant and algal growth. A majority of the excess nitrogen in either tank or pond culture systems originates as ammonia excreted by fish (Kibria *et al.*, 1998). The ammonia, as a waste product, is formed during the breakdown of proteins and excess amino acids not incorporated into tissues by the fish. Ammonia is toxic to fish and must be removed from the water. Fish excrete phosphorus in soluble and particulate forms. The soluble fraction called ortho phosphate (O-PO_4), is most available for plant growth. However, the main loading of phosphorus to the environment was reported to be *via* faecal pellets (Vielman *et al.*, 2000).

Hence, the aim of present investigation was to search an alternate protein source of plant origin that will not be only cost effective, but certainly would reduce excretion of nitrogenous wastes and total organic matter, possibly also of phosphorous, and alleviate pollution problems in intensive aqua-cultural systems.

Material and Methods

Specimens of *H. fossilis* were obtained from Ghazipur fish market, New Delhi, India. Specimens with mean body weight (5.2 g) were used in the studies. Fish were placed in the transparent glass aquaria (60 × 30 × 30 cm) kept in the laboratory where the temperature was maintained at 25±1°C and the lighting scheduled at 12h of light alternating with 12h of darkness. The fish were acclimatized for a minimum of 15 days prior to the initiation of experimental treatments. The water was renewed daily with chlorine free water.

Fishmeal and soybean seeds were used as the protein sources. Groundnut oil cake and rice bran were used as the basic feed ingredients. Soybean seeds were cleaned, autoclaved for half an hour at 121.6°C at 15 lb pressure to remove antinutritional factors (ANFs) such as trypsin inhibitors, haemoglutinins, lectins and phytic acid (Garg *et al.*, 2002). After oven drying at 60°C, it was ground into fine powder.

Groundnut oil cake, rice bran, fishmeal and processed soybean were finely ground to pass through 0.5 mm sieve. All the ingredients were subjected to proximate analysis following AOAC (1995) prior to the preparation of diets (Table 3.1). All the ingredients were mixed according to Table 3.2 and dough was made using distilled water. Thereafter, the dough was passed to a mechanical palletizer to obtain pellets (0.5 mm thick) which were dried in an oven and used in the studies for 80 days.

Table 3.1: Proximate Analysis (Per cent Dry Weight Basis) of the Ingredients Prior to the Preparation of Diets

Ingredients	*Proximate Composition (Per cent)*					
	Crude Protein	*Crude Fat*	*Crude Fiber*	*Total Ash*	*Nitrogen Free Extract*	*Gross Energy (KJg^{-1})*
Groundnut Oil Cake (GNOC)	35.266 ± 0.005	6.250 ± 0.003	6.000 ± 0.005	7.000 ± 0.000	45.496 ± 0.006	18.607 ± 0.003
Rice Bran (RB)	14.100 ± 0.005	10.066 ± 0.005	11.003 ± 0.005	20.556 ± 0.003	44.440 ± 0.005	14.906 ± 0.007
Fish Meal (FM)	42.926 ± 0.002	10.996 ± 0.003	3.493 ± 0.003	29.653 ± 0.003	12.930 ± 0.001	16.713 ± 0.000
Processed Soybean* (HPS)	43.733 ± 0.008	25.603 ± 0.000	4.496 ± 0.003	3.796 ± 0.000	22.370 ± 0.001	24.298 ± 0.001

* Raw soybeans were hydrothermically processed in an autoclave at 121°C at 15 lbs for 30 min. to remove anti-nutritional factors (ANFs) (Garg *et al.*, 2002).

\# All values are mean±S.E. of means of 3 observations.

The experimental diets were fed to duplicate groups of fish to satiation twice a day for a one week acclimatization period before starting the study. After this period, the fish were individually weighed and their initial weights recorded. The fish were then offered the test diets (1-10) twice a day (9:00h and 16:00h) to satiation, for 80 days. This period was considered enough to produce the effect of feeding on daily excretory pattern in the test species. Faeces were siphoned from culture aquaria every morning before fish feeding following the method of Spyridakis *et al.* (1989). In addition, about 20-50 per cent of the culture water was replaced daily with new, fresh, dechlorinated water. The pooled faecal samples were dried in an oven maintained at 60°C for subsequent analysis. Individual weight of the fish was recorded at the end of the experiment.

Analytical Techniques

The feed ingredients, experimental diets, faecal samples and fish carcass were analysed following the procedures stated in AOAC (1995). Chromide oxide levels in the diets as well as in the faecal samples were estimated spectrophotometrically according to the method of Furukawa and Tsukhara (1996). Live weight gain (g), growth percent gain, specific growth rate (SGR, per cent d^{-1}), protein efficiency ratio (PER) and gross conversion efficiency (GCE) were calculated using standard methods (Steffens, 1989). Apparent nutrient digestibility (APD) of the diets were calculated according to Cho *et al.* (1982) as follows:

$$\text{APD} = 100 - \frac{100 \times \text{per cent } Cr_2O_3 \text{ in diet} \times \text{per cent nutrient in faeces}}{\text{Per cent } Cr_2O_3 \text{ in faeces} \times \text{per cent nutrient in diet}}$$

Water Quality Parameters

On the last day of experiment offer the same feed to the fish in sufficient quantity so that the same is consumed, wait for 2 hours. Maintain a fixed level of water in each aquarium (say 30-40 L). Remove the excess of feed. The various water quality parameters like dissolved oxygen, pH, temperature,

Table 3.2: Ingredient Composition and Proximate Analysis (Per cent Dry Weight Basis) of 10 Compounded Diets (Diets 1-10) with Different Protein Levels (35, 38, 40, 42, 45 per cent)

	FM Based Diets (Protein per cent)					HPS Based Diets (Protein per cent)				
	1 (35)	2 (38)	3 (40)	4 (42)	5 (45)	6 (35)	7 (38)	8 (40)	9 (42)	10 (45)
Ingredients (Per cent)										
Groundnut Oil Cake (GNOC)[a]	60.0	60.0	60.0	60.0	60.0	60.0	60.0	60.0	60.0	57.9
Rice Bran (RB)[b]	18.0	10.0	5.0	1	–	17.5	9.8	7.6	0.7	–
Fish Meal (FM)[c]	15.0	23.0	28.0	32.0	35.0	–	–	–	–	–
Processed Soybean (HPS)[d]	–	–	–	–	–	15.5	23.2	28.4	32.3	35.1
Binder[e]	5.0	5.0	5.0	5.0	5.0	5.0	5.0	5.0	5.0	5.0
Chromic Oxide (Cr_2O_3)[f]	1.0	1.0	1.0	1.0	1.0	1.0	1.0	1.0	1.0	1.0
Mineral premix and amino acids (MPA)[g]	1.0	1.0	1.0	1.0	1.0	1.0	1.0	1.0	1.0	1.0
Proximate Analysis (Per cent)										
Crude Protein	35.00	38.00	40.25	42.00	45.50	35.00	38.5	40.25	42	45.5
Crude Fat	5.5	6.3	7	9	8.5	7	10.5	9.5	11	12
Crude Fiber	3.5	4	3.5	5	4.5	6.5	6	7.25	6	5
Ash	9.15	8.5	7.3	5.95	6.65	9.05	8.45	6.5	6:8	7.15
Nitrogen Free Extract (NFE)	46.85	43.2	42.95	38.05	34.85	42.45	36.55	36.5	35	30.35
Gross Energy KJg^{-1}	18.48	18.88	19.65	20.01	20.09	18.32	19.52	19.53	20.28	20.71

a, b: Used as basic feed ingredients; c: Used as the reference protein source of animal origin; d: Used as the main protein source of plant origin; e: Used is Carboxyl methyl cellulose to make the diets water stable; f: Used as an external indigestible marker for estimating apparent digestibility; g: Added to supplement the diets with minerals and amino acids.

Each Kg contains Copper: 312mg; Cobalt: 45mg; Magnesium: 2.114g; Iron: 979mg; Zinc: 2.13g; Iodine: 156mg; DL-Methionine: 1.92g; L-lysine mono hydrochloride: 4.4g; Calcium 30 per cent and Phosphorous–8.25 per cent.

conductivity, free carbondioxide, total alkalinity and total hardness of aquaria water were analyzed following APHA (1998). Start collecting water samples from each aquarium in replicate of 2 for the determination of ammonical nitrogen (NH_4-N) and ortho-phosphate (O-PO_4) following APHA (1998) to see the influence of compounded feeds on pollution status of receiving water in the aquaria.

Calculate the excretory levels of NH_4-N and O-PO_4 in treated water as follows:

$$\text{NH}_4\text{-N excretion (mg/100g BW of fish)} = \frac{\text{NH}_4\text{-N (mg l}^{-1}\text{) in aquarium water}}{\text{Fish weight (mg) per L of water}}$$

$$\text{O-PO}_4\text{ excretion (mg/100g BW of fish)} = \frac{\text{O-PO}_4\text{ (mg l}^{-1}\text{) in aquarium water}}{\text{Fish weight (mg) per L of water}}$$

Statistical Analysis

Data was analysed following ANOVA, Duncan Multiple Range Test (Duncan, 1955) and Multivariate Analysis (Prein *et al.*, 1993) at 5 per cent probability level. Group means were compared by student 't' test.

Results and Discussions

Water Quality Parameters

The effects of 10 different diets are very clearly reflected on the physico-chemical characteristics of aquaria water (Table 3.3 and 3.4).

The tables showed that DO fluctuated between 5.25 to 5.70 mg/l, conductivity between 0.48 to 0.52 micro mhos/cm, free CO_2 between 16.00 to 17.40 mg/l, total alkalinity between 214.00 to 254.00 mg/l and total hardness between 212.00 to 228.00 mg/l. The pH remained alkaline (7.4 to 7.8). The water temperature fluctuated between 27.0 to 28.0°C.

Although DO levels, remained at optimum levels, but low DO values in aquaria where the fish were fed on 40.25 per cent dietary protein (HPS based diets) clearly indicated its utilization by the growing fish (Kalla *et al., 2003;* Jindal, 2008).

Post-prandial Excretory Levels of Ammonical Nitrogen (NH_4-N) and Ortho-phosphate (O-PO_4)

A significant ($P<0.05$) decrease in NH_4-N excretion and O-PO_4 production was observed in the aquaria waters up to the dietary protein levels 40.25 per cent, irrespective of the protein source (Table 3.3 and 3.4). But as the protein level increased above 40.25 per cent, increase in NH_4-N excretion and O-PO_4 production was observed. These studies showed that digestibility of protein in the test species is only up to 40.25 per cent protein level and above this level, they will starts polluting the aquaria water resulting in poor growth of fish. These results are in agreement with those of Kalla *et al., 2003* and Jindal, 2008.

The results further indicated that the level of NH_4-N excretion and O-PO_4 production were found significantly ($P<0.05$) low in fish fed on HPS based diets (plant proteins) as compared to FM based diets (animal protein). Reduction in NH_4-N excretion and O-PO_4 production with the use of HPS in the diets have been reported by many other workers (Singh *et al.*, 2003; Kalla and Garg, 2004; Jindal, 2008).

Table 3.3: Water Quality Parameters of Different Aquariums Stocked with *H. fossilis* Fingerlings Fed on FM Based Diets (Diets 1–5) at Different Dietary Protein Levels

Parameters	Diet No.				
	1 (35)	2 (38)	3 (40)	4 (42)	5 (45)
Dissolved oxygen (DO) mg/l	5.45± 0.001	5.65± 0.003	5.30± 0.002	5.70± 0.008	5.46± 0.004
pH	7.50	7.60	7.60	7.80	7.70
Water temperature (°C)	28.00	28.00	28.00	27.50	27.50
Conductivity micro (μ) mhos cm^{-1}	0.48± 0.005	0.49± 0.003	0.49± 0.000	0.50± 0.002	0.52± 0.007
Free Carbon dioxide (Free CO_2) mg/l	16.00± 0.000	16.40± 0.002	16.20± 0.001	16.20± 0.004	16.00± 0.002
Total alkalinity (mg/l)	215.0± 0.002	225.0± 0.000	215.0± 0.001	216.0±0.003	214.0± 0.001
Total hardness (mg/l)	213.0± 0.005	212.0± 0.004	219.0± 0.001	218.0± 0.003	228.0± 0.002
Ammonical nitrogen (NH_4-N) excretion (mg/100BW of fish)	1.935± 0.015	1.845± 0.045	1.245±0.015	1.465± 0.035	1.885± 0.015
Ortho-phosphate (O-PO_4) excretion (mg/100BW of fish)	1.740± 0.020	1.700± 0.030	0.950± 0.020	1.320± 0.010	1.540± 0.002

*: All values are mean±S.E. of mean of 3 observations.

Table 3.4: Water Quality Parameters of Different Aquariums Stocked with *H. fossilis* Fingerlings Fed on HPS Based Diets (diets 6–10) at Different Dietary Protein Levels

Parameters	Diet No.				
	6 (35)	7 (38)	8 (40)	9 (42)	10 (45)
Dissolved oxygen (DO) mg/l	5.35± 0.002	5.40± 0.003	5.25± 0.001	5.55± 0.002	5.35± 0.001
pH	7.40	7.40	7.60	7.80	7.80
Water temperature (°C)	27.50	27.00	28.00	28.00	28.00
Conductivity micro (μ) mhos cm^{-1}	0.49± 0.001	0.50± 0.004	0.48± 0.002	0.51± 0.000	0.50± 0.001
Free carbondioxide (Free CO_2) mg/l	16.40± 0.000	16.80± 0.001	17.40± 0.005	17.20± 0.001	17.20± 0.002
Total alkalinity (mg/l)	222.0± 0.005	223.0± 0.003	242.0± 0.001	248.0± 0.002	254.0± 0.003
Total hardness (mg/l)	228.0± 0.002	226.0± 0.005	228.0± 0.002	219.0± 0.001	221.0± 0.005
Ammonical nitrogen (NH_4-N) excretion (mg/100BW of fish)	1.830± 0.028	1.775± 0.021	1.350± 0.028	1.620± 0.042	1.145± 0.035
Ortho-phosphate (O-PO_4) excretion (mg/100BW of fish)	1.555± 0.049	1.450± 0.028	1.140± 0.020	1.270± 0.031	1.335± 0.035

*: All values are mean±S.E. of mean of 3 observations.

These results further showed that plant protein based diets excrete less ammonia and phosphorous and thus less pollute the water.

In the present studies, maximum NH_4-N excretion occurred after 8h post feeding (Figure 3.1) and maximum O-PO_4 production after 6h post feeding (Figure 3.2) irrespective of the protein source. Jindal and Garg, 2005; Jindal *et al.*, 2007a,b; Robinson and Menghe, 2007 and Jindal, 2008 also reported the NH_4-N excretion and O-PO_4 production after 6-9h of post feeding.

Growth Parameters

Survival in different treatments were high and varied between 84 to 93 per cent (Table 3.5 and 3.6). Fingerlings in different treatments were fed on one of the 10 formulated diets containing protein levels ranging from 35.0 to 45.5 per cent. ANOVA revealed that a significant ($P<0.05$) high live weight gain, per cent weight gain and SGR were observed in the fingerlings fed on diets containing 40.25 per cent dietary protein, irrespective of the protein source. On the other hand, FCR values at 40.25 per cent dietary protein level remained significantly ($P<0.05$) low in comparison to other dietary treatments. The data further indicates that an increase in dietary protein contents (beyond 40.25 per cent) not only repressed growth performance but also significantly ($P<0.05$) increased FCR values. The data further

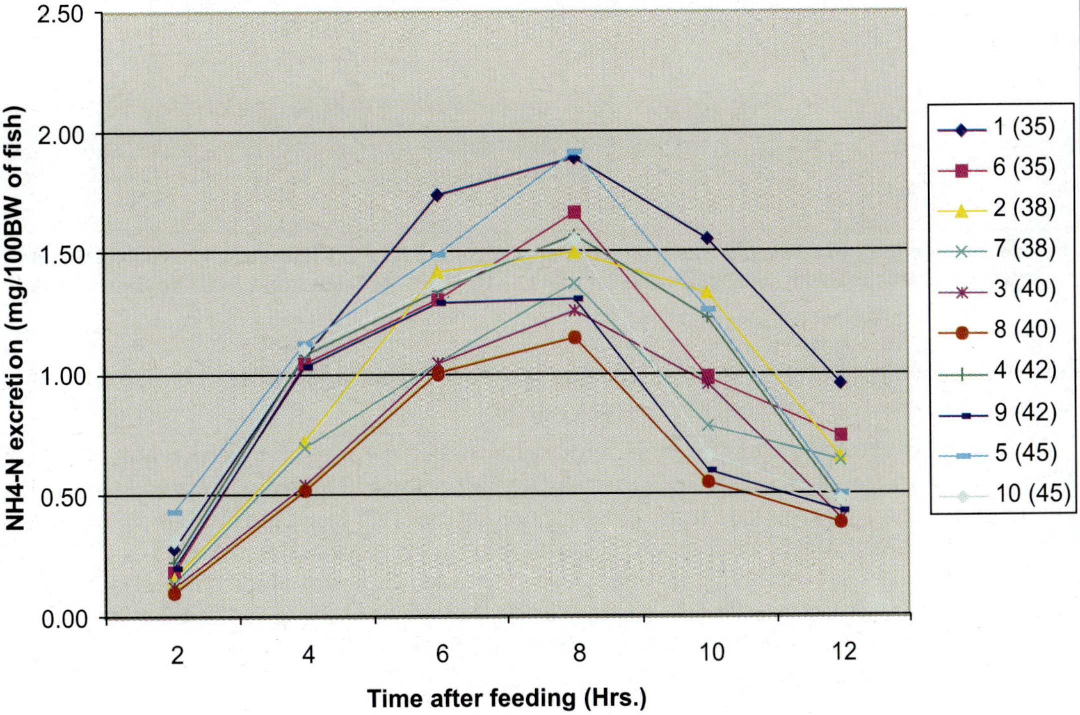

Fig. 1

Figure 3.1: Comparative Diurnal Pattern of Ammonical Nitrogen (NH_4-N) Excretion in Treated Waters of Aquariums Stocked with *H. fossilis* Fed on Diets 1–10 (1–5–FM based and 6–10 Soybean based)

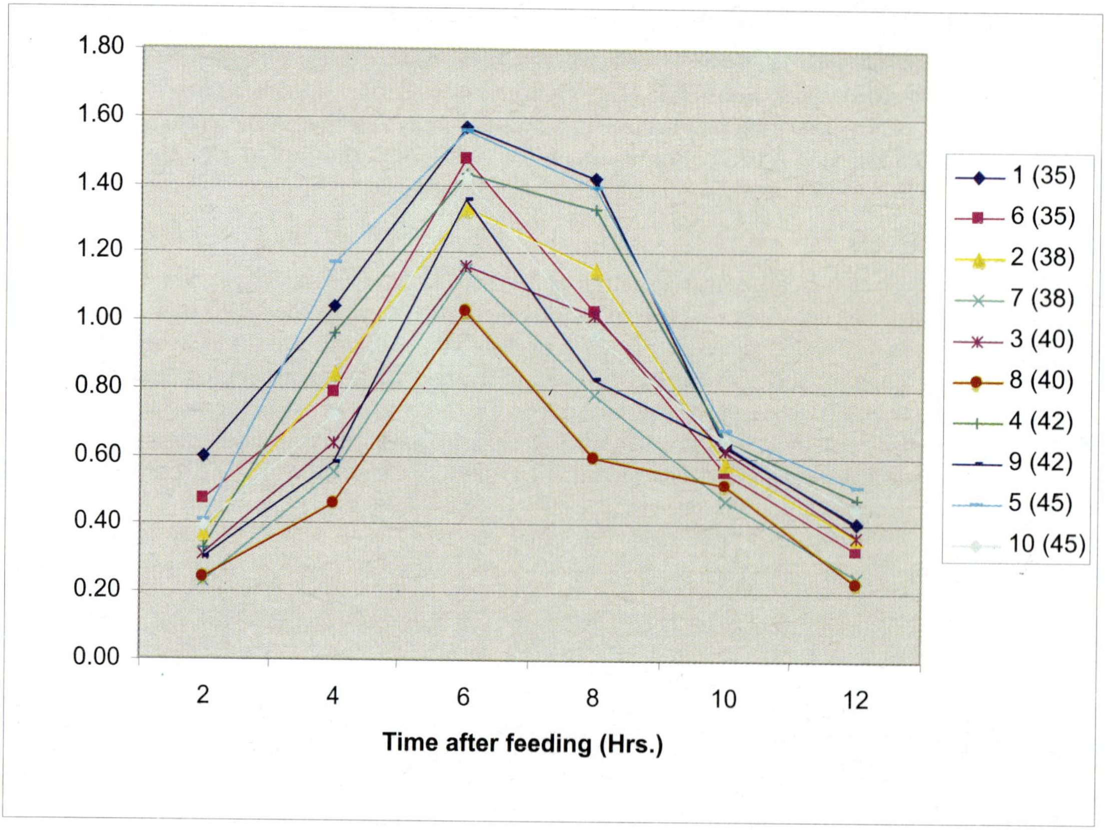

Figure 3.2: Comparative Diurnal Pattern of Ortho-phosphate (O-PO_4) Excretion in Treated Waters of Aquariums Stocked with *H. fossilis* fed on diets 1–10 (1–5–FM based and 6–10 Soybean based)

showed a slight high growth performance at 40.25 per cent dietary protein from HPS based diets may be attributed to the high digestibility of plant protein.

Low growth at other dietary treatments may indicated that optimum protein requirements of *H. fossilis* fingerlings is about 40.25 per cent. Feeding the fingerlings on high dietary protein level (45 per cent) not only repressed growth performance, but also deteriorated water quality as evident from high NH_4-N and O-PO_4 excretion. Protein levels above optimum requirements may results in decreased growth rates because of a reduction in dietary energy available for growth due to energy required to deaminate and excrete excess absorbed amino acids (Kalla and Garg, 2004; Jindal and Garg, 2005; Jindal *et al.*, 2007a,b; Robinson and Menghe, 2007 and Jindal, 2008).

The studies further showed that growth and digestibility parameters were found to be negatively correlated with NH_4–N and O-PO_4 excretion. This is also the reason for the better growth of fish fed on diet 4 than on diet 3. A comparison of weight gain in fish groups fed on FM based and HPS based diets is shown in Figure 3.3.

Table 3.5: Effect of Five Dietary Protein Levels on Growth Performance, Digestibility and Nutrient Retention in *H. fossilis* Fingerlings by Using FM as the Main Protein Source

Parameters	*Diets (Per cent Protein)*				
	1 (35)	*2 (38)*	*3 (40)*	*4 (42)*	*5 (45)*
Survival (Per cent)	85	84	93	90	88
Live Weight gain (g)	$11.775^{A}\pm0.725$	$14.985^{B}\pm0.435$	$17.210^{C}\pm0.650$	$15.100^{B}\pm0.090$	$14.920^{B}\pm0.070$
Fish length gain (cm)	$6.850^{A}\pm0.350$	$9.200^{B}\pm0.200$	$11.600^{C}\pm0.000$	$9.700^{B}\pm0.200$	$9.050^{B}\pm0.050$
Growth per cent gain in body wt	$134.917^{A}\pm12.141$	$183.106^{BCD}\pm0.837$	$200.643^{BC}\pm2.082$	$170.634^{BD}\pm1.283$	$168.896^{BD}\pm1.597$
Growth/day per cent body wt	$1.023^{A}\pm0.072$	$1.194^{B}\pm0.002$	$1.252^{B}\pm0.006$	$1.144^{B}\pm0.011$	$1.148^{B}\pm0.003$
Specific growth rate SGR (Per cent d^{-1})	$0.448^{A}\pm0.015$	$0.564^{BD}\pm0.002$	$0.597^{C}\pm0.004$	$0.536^{BDE}\pm0.006$	$0.535^{DE}\pm0.005$
Food Consumption/day per cent BW	$4.268^{AC}\pm0.062$	$4.011^{B}\pm0.013$	$4.172^{AC}\pm0.017$	$4.216^{AC}\pm0.020$	$4.132^{BC}\pm0.033$
Feed Conversion ratio (FCR)	$4.194^{A}\pm0.358$	$3.356^{B}\pm0.018$	$3.333^{B}\pm0.003$	$3.684^{AB}\pm0.018$	$3.599^{B}\pm0.022$
Protein efficiency ratio (PER)	$0.336^{A}\pm0.028$	$0.394^{BC}\pm0.012$	$0.428^{BC}\pm0.016$	$0.359^{AB}\pm0.002$	$0.327^{A}\pm0.002$
Gross conversion ratio (GCE)	$0.240^{A}\pm0.021$	$0.297^{B}\pm0.002$	$0.300^{B}\pm0.001$	$0.272^{AB}\pm0.002$	$0.277^{B}\pm0.002$
Apparent protein digestibility (APD) per cent	$80.000^{A}\pm0.000$	$81.430^{B}\pm0.000$	$85.250^{C}\pm0.250$	$84.400^{D}\pm0.100$	$82.685^{E}\pm0.175$

BW: Body Weight.

Duration of experiment: 80 days

Mean with the same letter in the same column are not significantly ($P>0.05$) different.

All values are Mean±S.E. of mean of three observations.

Data was analyzed by Duncan Multiple Range test.

Table 3.6: Effect of Five Dietary Protein Levels on Growth Performance, Digestibility and Nutrient Retention in *H. fossilis* Fingerlings by Using HPS as the Main Protein Source

Parameters	*Diets (Per cent Protein)*				
	6 (35)	*7 (38)*	*8 (40)*	*9 (42)*	*10 (45)*
Survival (Per cent)	90	88	92	80	85
Live Weight gain (g)	$14.400^{A}\pm0.510$	$17.100^{B}\pm0.290$	$21.905^{C}\pm0.045$	$18.625^{B}\pm0.325$	17.930 ± 0.760
Fish length gain (cm)	$10.050^{A}\pm0.050$	$11.600^{B}\pm0.200$	$14.600^{C}\pm0.100$	$11.900^{B}\pm0.600$	$11.100^{AB}\pm0.400$
Growth per cent gain in body wt	$165.139^{A}\pm7.830$	$205.138^{B}\pm8.761$	$269.511^{C}\pm4.864$	$216.664^{B}\pm6.802$	$224.445^{B}\pm4.880$
Growth/day per cent body wt	$1.148^{A}\pm0.047$	$1.265^{B}\pm0.027$	$1.434^{C}\pm0.011$	$1.299^{B}\pm0.019$	$1.323^{B}\pm0.015$
Specific growth rate SGR (Per cent d^{-1})	$0.271^{A}\pm0.036$	$0.389^{B}\pm0.033$	$0.537^{C}\pm0.013$	$0.419^{B}\pm0.024$	$0.438^{B}\pm0.017$
Food Consumption/day per cent BW	$4.075^{A}\pm0.031$	$3.949^{A}\pm0.079$	$4.150^{A}\pm0.168$	4.046 ± 0.076	$4.133^{A}\pm0.053$
Feed Conversion ratio (FCR)	$3.554^{A}\pm0.168$	$3.123^{B}\pm0.111$	$2.893^{B}\pm0.105$	$3.114^{B}\pm0.089$	$3.127^{B}\pm0.060$
Protein efficiency ratio (PER)	$0.411^{AB}\pm0.015$	$0.449^{AB}\pm0.007$	$0.544^{C}\pm0.001$	$0.443^{AB}\pm0.007$	$0.394^{A}\pm0.017$
Gross conversion ratio (GCE)	$0.282^{AB}\pm0.013$	$0.321^{ABC}\pm0.012$	$0.346^{BC}\pm0.012$	$0.322^{BC}\pm0.009$	$0.320^{ABC}\pm0.006$
Apparent protein digestibility (APD) per cent	$81.375^{A}\pm0.531$	$82.785^{B}\pm0.160$	$87.275^{C}\pm0.318$	$86.125^{D}\pm0.176$	$85.420^{D}\pm0.424$

BW: Body Weight.

Duration of experiment: 80 days

Mean with the same letter in the same column are not significantly ($P>0.05$) different.

All values are Mean±S.E. of mean of three observations.

Data was analyzed by Duncan Multiple Range test.

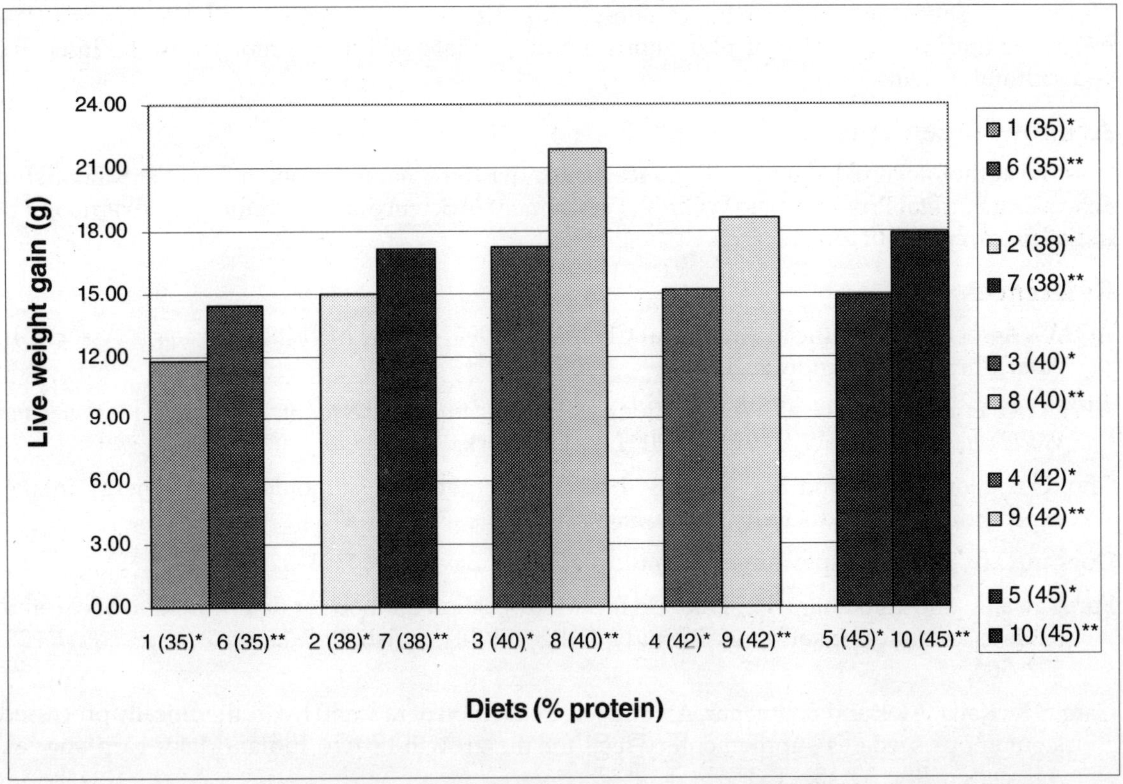

Figure 3.3: Comparison of Feeding 10 Diets Containing 5 Different Dietary Protein Levels (Diets 1–5: Fishmeal* and 6–10 Processed Soybean) on Live Weight Gain in *H. fossilis* Fingerlings**

Conclusion

Fish have the ability to handle protein in excess of that needed for growth and maintenance by deaminizing amino acid bronchially and excreting ammonia. Present studies on *H. fossilis*, thus, have established that when the protein levels in the diets exceeds the limits of digestibility (above 40.25 per cent) are deaminized and are excreted as ammonia in the aquaria water, which sometimes may be stressful for the fish especially in some impoundments and impedes growth.

Further, the use of plant proteins make the aquaculture feed more environmental friendly as the metabolism of such feeds results in the lower excretion of nitrogen, phosphorous and other organic wastes in the environment. Higher the ingestion of proteins by the fish, more ammonia, urea and phosphorous are excreted in the environment, therefore, determination of nutrient budgets and daily pattern of excretion of metabolites by the fish is important for evaluating the potential waste load of the fish farm effluents. Studies have revealed that the use of plant proteins in fish feeds not only reduces the cost of feed formulation but also reduces the post prandial excretion of ammonia and phosphorous in the treated waters.

The results of this study clearly demonstrated that HPS supplemented with MPA can be recommended as a replacement of FM based diet for *H. fossilis*, up to the protein level of 40.25 per cent.

This not only saves total feed cost, but certainly would also reduce excretion of nitrogenous and total organic matter, possibly also of phosphorous and alleviate pollution problems in the intensive aquacultural systems.

Acknowledgements

The author acknowledges funding received under the scheme "Women Scientist Scholarship Scheme for Societal Programmes (WOS-B), Department of Science and Technology, Government of India" for carrying out this research.

References

AOAC (Association of Official Analytical Chemists), 1995. *Official Methods of Analysis*. Assoc. Off. Anal. Chem. Washington, Sc, USA.

APHA (American Public Health Association), 1998. *Standard Methods for the Examination of Water and Wastewater*. APHA, AWWA, EPFC, 20th Ed., New York.

Cho, C.Y., Slonger, S.J. and Bayley, H.S., 1982. Bioenergetics of salmonid fishes. Energy intake, expenditure and productivity. *Comp. Biochem. Physiol.*, 73B: 25–41.

Duncan, D.B., 1955. Multiple range and multiple F-tests. *Biometrics*, 11: 1–42.

Furukawa, A. and Tsukahara, H., 1966. On the acid digestion method for determination of chromic oxide as an indicator substance in study of digestibility in fish. *Bull. Japanese Soc. Sci. Ferti.*, 32: 502–506.

Garg, S.K., Kalla, Alok and Bhatnagar, Anita, 2002. Evaluation of raw and hydrothermically processed leguminous seeds as supplementary feed for the growth of two Indian Major carp species. *Aquaculture Res.*, 33: 151–163.

Jindal, M., Garg, S.K., Yadava, N.K. and Gupta, R.K., 2007b. Effect of replacement of fishmeal with processed soybean on growth performance and nutrient retention in *Channa punctatus* (Bloch.) fingerlings. In: *Livestock Research for Rural Development*, Volume 19, Article #165. Retrieved from http://www.cipav.org.co/lrrd/lrrd19/11/jind19165.htm.

Jindal, M., Garg, S.K. and Yadava, N.K., 2007a. Effect of replacement of fishmeal with dietary protein sources of plant origin on the growth performance and nutrient retention in the fingerlings of *Channa punctatus* (Bloch.) for sustainable aquaculture. *Pb. Univ. Res. J. (Sci.)*, 57 (in press).

Jindal, Meenakshi and Garg, S.K., 2005. Effect of replacement of fishmeal with defatted canola on growth performance and nutrient retention in the fingerlings of *channa punctatus* (Bloch.). *Pb. Univ. Res. J. (Sci.)*, 55: 183–189.

Jindal, Meenakshi, 2008. Studies on protein requirements of catfish *Clarias batrachus* fingerlings for sustainable aquaculture. In: *National Seminar on Technical Advances in Environment Management and Applied Zoology*, Jan. 23–25, Department of Zoology, Kurukshetra University, Kurukshetra, pp. 59–62.

Kalla, Alok and Garg, S.K., 2004. Use of plant proteins in supplementary diets for sustainable aquaculture. In: *National Workshop on Rational Use of Water Resources for Aquaculture*, (Eds.) S.K. Garg and K.L. Jain, March 18–19, Hisar, India, pp. 31–47.

Kalla, Alok, Garg, S.K. and Kaushil, C.P., 2003. Effect of dietary protein source on growth, digestibility and body composition in the fingerlings of *Cirrhinus* mrigala (Ham.) In: *Proceedings of 3rd Interaction*

Workshop on Fish Production Using Brackish Water in Arid Eco-system, (Eds.) S.K. Garg and A.R.T. Arasu, December 17–18, Hisar, India, pp. 139–145.

Kibria, G., Nugegoda, D., Fairclough, R. and Lam, P., 1998. Can nitrogen pollution from aquaculture be reduced? *NAGA, ICLARM*, 21: 17–25.

Prein, M., Hulata, G. and Pauly, D., 1993. On the use of multivariate statistical methods in aquacultural research. In: *Multivariate Methods in Aquacultural Research: Case Studies of Tilatias in Experimental and Commercial Systems*, (Eds.) M. Prein, G. Hulata and D. Pauly. ICLARM. Stud. Rev., 10: 1–12.

Robinson, Edwin H. and Menghe, H.Li., 2007. *Catfish Protein Nutrition (Revised)*. Bulletin 1153. Office of Agricultural Communications, Mississippi State University, USA.

Singh, K., Garg, S.K., Alok, Kalla and Bhatnagar, Anita, 2003. Oil cakes as protein source in supplementary diets for the growth of *Cirrhinus mrigala* (Ham.) fingerlings. Laboratory and field studies. *Bioresource Technology*, 86: 283–291.

Spyridakes, P., Metailler, R., Gabandan, J. and Riaza, A., 1989. Studies on nutrient digestibility in European sea bass *Dicentrarchus labrax* I. Methodical aspects concerning faeces collection. *Aquaculture*, 77: 61–70.

Steffens, W., 1989. *Principles of Fish Nutrition*. Ellis Horwood, Chichester.

Vielman, J., Makinen, T., Ekholm, P. and Koskela, J., 2000. Influence of dietary soybean and phytase levels on performance and body composition of large rainbow trout (*Oncorhynchus mykiss*) and algal availability of phosphorous load. *Aquaculture*, 183: 349–362.

Chapter 4

Bactericides from Actinobacteria Isolated from the Sediments of Shrimp Pond

☆ *K. Sivakumar, Maloy Kumar Sahu, V. Arul, Prashant Kumar, S. Raja, T. Thangaradjou and L. Kannan*

ABSTRACT

A total number of 20 strains of actinobacteria were isolated from the sediments of a south India shrimp pond and tested for their antibacterial activity against some potential human pathogens such as *Bacillus* sp., *Klebsiella* sp., *Proteus* sp., *Pseudomonas* sp. and *Streptococcus* sp., by cross-streak method. Out of 20 strains, 9 strains (45 per cent) were found to be inhibitory to one or more human bacterial pathogens at varying levels. Among them, 25 per cent strains were effective against *Klebsiella* sp., *Proteus* sp. and *Pseudomonas* sp. Another 15 per cent strains were found antagonistic against *Bacillus* sp. and remaining 5 per cent of the strains were effective against *Streptococcus* sp. The effective 9 strains were further examined in agar disc method and among them 3 strains (PS-4, PS-9 and PS-14) showed good activity against all the tested pathogens and the strains were tentatively identified as *Streptomyces moderatus* (PS-4), *Streptomyces violaceus* (PS-9) and *Streptomyces gibsonil* (PS-14). Further studies are continued to isolate, purify and characterize the bactericides from these antagonistic actinobacteria and to test their possible use as the alternative chemotherapeutic drugs on a commercial scale.

Introduction

The screening of microbial natural products continues to represent an important route for the discovery of novel chemicals for development of new therapeutic agents and for evaluation of the potential of lesser-known and/or new bacterial taxa (Kurtboke and Wildman, 1998). It has been estimated that approximately two-thirds of the thousands of naturally occurring antibiotics have been isolated from actinobacteria (Takizawa *et al.*, 1993). Indeed, the *Streptomyces* species produce about 75 per cent of commercially and medicinally useful antibiotics (Miyadoh, 1993) and most of these antibiotics are coming from the terrestrial actinobacteria. But, marine actinobacteria would be very important sources for the discovery of new metabolites (Okami, 1986). The actinobacteria found in the marine and coastal ecosystems may be viewed as a rich gene pool possibly containing isolates capable of producing useful metabolites (Okami, 1986; Goodfellow and Haynes, 1984). In recent years, there has been a growing awareness of the potential value of marine sediments as sources of actinobacteria that produce useful metabolic products (Goodfellow and Haynes, 1984). However, only few attempts have been made to isolate marine actinobacteria for their antagonistic activity (Balagurunathan, 1992; Balagurunathan and Subramanian, 1998; Dhevendaran and Annie 1999; Patil *et al.*, 2001; Sahu *et al.*, 2004; Sivakumar *et al.*, 2005a,b,c; Sahu *et al.*, 2005a,b; Umamaheswary *et al.*, 2005; Muthurayar *et al.*, 2006; Sivakumar *et al.*, 2006; Sahu *et al.*, 2006). From the available literature, it is found that most of the works on actinobacteria were carried out from different marine environs but there is no work on the actinobacteria isolated from shrimp ponds. Hence, the present study was undertaken with the aim of isolating actinobacteria, antagonistic to human bacterial pathogens from a south Indian shrimp pond.

Material and Methods

Collection of Samples

Sediment samples were collected from an estuarine shrimp pond (Lat. 11°28′ 53.7″N and Long. 79°45′ 31.1″E), located opposite to the Vellar estuary, southeast coast of India by inserting a sterilized polyvinyl corer (10 cm) into the sediments. The centre portion of the 2 cm sediment sample was taken out with the help of a sterile spatula. This sample was then transferred to a sterile polyethylene bag and transported immediately to the laboratory.

Isolation of Actinobacteria

The sediment samples collected were air-dried aseptically. After a week, they were incubated at 55°C for 5 min (Waksman and Lechvalier, 1962). Then, 10-fold serial dilutions of the samples were prepared using filtered and sterilized seawater. One ml of the serially diluted sample was plated in Kuster's Agar medium prepared in 50 per cent seawater to enhance the isolation of marine actinobacteria (Sivakumar *et al.*, 2005a). pH of the selected medium used was adjusted to 7.2. To prevent the fungal and bacterial contaminations, cycloheximide (100mg/l) and nalidixic acid (20 mg/l) were added to the medium (Kathiresan *et al.*, 2005). The petri plates were then incubated at 28±2°C and the colonies were observed from 5th day onwards for 1 month (Sivakumar *et al.*, 2005a). Strains of actinobacteria were picked out and purified by repeated streaking on yeast extract-malt extract agar (ISP-2) medium. The pure cultures of the actinobacteria were transferred to ISP-2 slants and preserved at 4±2°C.

Screening of Actinomycete Strains for Antibacterial Activity

Primary Screening

Antibacterial activity of isolated actinobacteria against *Bacillus* sp., *Klebsiella* sp., *Proteus* sp., *Pseudomonas* sp. and *Streptococcus* sp. was studied. The antibacterial activity was tested, using the cross streak method (Waksman and Lechevalier, 1962). Single streak of the actinomycete was made on the surface of the modified nutrient agar (Sivakumar *et al.*, 2005a) and incubated at room temperature

(28±2°C). After observing a good ribbon-like growth of the actinomycete on the petriplates, the pathogen was streaked at right angles to the original streak of the actinomycete and incubated at 28±2°C. The inhibition zone was measured after 24 and 48 h. A control plate was maintained without inoculating the actinomycete, to assess the normal growth of bacteria. From this screening, strains of potential antagonistic actinobacteria were selected.

Secondary Screening

The selected strains were further tested in secondary screening by disc method (Bauer *et al.*, 1966). The partially purified extract obtained by the evaporation of the ethyl acetate extract was dissolved in 1 ml 0.2M phosphate buffer (pH 7.0). Then 100ml of it was loaded into well bored and text organism (0.5 McFarland turbidity standard) swabbed Muller Hinton agar plates. The plates were incubated at 37°C for 18-24 h and examined. The diameter of the zones of complete inhibition was measured to the nearest whole millimeter.

Characterization of Antagonistic Actinobacteria

The genus level identification was made for the strains which showed good antagonistic activity, using cell wall composition analysis and micromorphological studies (Lechevalier and Lechevalier 1970). Species level identification of these strains was based on the methods described by Shirling and Gottlieb (1966), Key of Nonomura (1974) and Bergey's Manual of Determinative Bacteriology (Buchanan and Gibbons, 1974).

Results and Discussion

Actinobacteria in Shrimp Pond Sediments

In the present study, 22 actinomycete strains were isolated from the sediments of shrimp pond. As the shrimp ponds are located on the banks of the Vellar estuary and waters are taken from the estuary and there is every possibility for the actinobacteria to enter into the pond and, could settle and regenerate in the pond sediments. So the isolated actinobacteria might have originated from the estuary. Higher density of actinobacteria enumerated isolated from shrimp sediments could be due to the continuous availability of substrate and nutrients in the form of unconsumed feed, shrimp excreta, plankton and other organic and inorganic matters at the pond bottom. It is worth mentioning that it has been reported that about 35 per cent of shrimp feed supplied is either leached or unconsumed (15 per cent) or egested (20 per cent) as faces (Primavera, 1994). Maclean *et al.* (1994), Hameed (1993), Sharmila *et al.* (1996), Otta *et al.* (1999), Janakiram *et al.* (2000) and Dalmin *et al.* (2002) have also recorded more numbers of bacterial isolates from shrimp pond sediment.

Antibacterial Activity

In the present study, 20 strains were isolated and among them 9 strains (45 per cent) were found to be inhibitory to one or more human bacterial pathogens at varying levels. Out of the 20 strains, 25 per cent strains were effective against *Klebsiella* sp., *Proteus* sp. and *Pseudomonas* sp. Moderate percentage of strains (15 per cent) was found antagonistic against *Bacillus* sp. Only 5 per cent of the strains were effective against *Streptococcus* sp. Earlier, a similar result was observed, in which, it was found that about 41.67 per cent of the actinomycete isolates were inhibitory to one or more human bacterial pathogens at varying levels (Sivakumar *et al.*, 2005a). However, in another study, it has been reported that only 12.5 per cent of actinobacteria cultures isolated from the sediments of the Vellar estuary were antagonistic to various human bacterial pathogens (Sahu *et al.*, 2006). These reports suggest that sediments can be the potent source of antibacterial compounds.

Of the effective isolates, only 4 strains were selected for the secondary screening, based on their wide antagonistic activity against all the five pathogenic bacteria tested. The results are given in Table 4.1. Among the four strains, strains PS-4 and PS-9 exhibited inhibition zone against all the tested bacterial pathogens whereas strain PS-14 showed inhibition to all the pathogens except *Bacillus* sp. The strain PS11 did not exhibit any inhibition against any tested bacterial pathogens. Hence, the strains PS-4, PS-9 and PS-14 were considered as the potent source for antibiotics and were selected for identification.

Table 4.1: Antagonistic Activity of Four Strains of Actinobacteria, Isolated from the Sediments of Shrimp Pond

Strain No.	*Activity Against Human Bacterial Pathogens (Inhibition zone in mm)*				
	Bacillus sp.	*Klebsiella sp.*	*Proteus sp.*	*Pseudomonas sp.*	*Streptococcus sp.*
PS–4	16	14	10	8	11
PS–9	11	9	7	13	6
PS–11	–	–	–	–	–
PS–14	–	13	11	15	9

Characterization of Antagonistic Actinobacteria

All the three strains *viz.* PS-4, PS-9 and PS-14 possess LL-Diaminopimelic Acid and contain glycine in their cell wall (Table 4.2). Presence of LL-Diaminopimelic Acid along with glycine indicates the cell wall chemotype–1. The strains with chemotype–1 do not have characteristics pattern of sugars (Lechevalier and Lechevalier, 1970).

Table 4.2: Cell Wall Amino Acids and Whole Cell Sugars of Three Strains of Actinobacteria

Strain No.	*LL-DAP*	*Meso-DAP*	*Glysine*	*Whole Cell Sugars*	*Wall Type*
PS–4	+	–	+	Nd	1
PS–9	+	–	+	Nd	1
PS–14	+	–	+	Nd	1

+: Positive; –: Negative; Nd: Not determined.

The representatives belonging to the wall type 1 are *Streptomyces, Stretoverticillium, Chainia, Actinopycnidium, Actinosporangium, Elytrosporangium, Microellobosporia, Sporichthya* and *Intrasporangium* (Lechevalier and Lechevalier, 1970). The micromorphological observations of the strains PS-4, PS-9 and PS-14 reveal that all these belong to the genus *Streptomyces*. The morphological, micromorphological, physiological and biochemical characteristics of antagonistic strains (PS-4, PS-9 and PS-14) tested in the present study are given in Tables 4.3 and 4.4. These characteristics were compared with those *Streptomyces* species given in the key of Nonomura (1974) and described in the Bergey's Manual of Determinative Bacteriology (Buchanan and Gibbons, 1974).

The strain PS-4 was weak in the production of soluble pigment. Except this, all the other characters of the strain PS-4 were y similar to that of *Streptomyces moderatus* and hence the strain PS-4 has been tentatively identified as *S. moderatus* (Tables 4.3 and 4.4). Similarly, the strain PS-9 was weak in utilizing carbon compounds *viz.* arabinose and xylose (Table 4.4). Except this, all the other characteristics properties are the same in PS-9 and *Streptomyces violaceus*. Therefore, the strain PS-9 has been tentatively identified as *S. violaceus*. In the strain PS-14, except for the difference in the utilization of inositol (Table 4.4), all the other characteristics are exactly similar to that of *Streptomyces gibsonil* and this strain has been tentatively identified as *S. gibsonil*.

Table 4. 3: Morphological and Micromorphological Characteristics of Three Strains of Antagonistic Actinobacteria

Strain No.	*Species to which Assigned*	*Aerial Mass Colour*	*Melanoid Pigment*	*Reverse Side Pigment*	*Soluble Pigment*	*Spore Chain*
PS-4	*Streptomyces moderatus*	White	–	+	–	Rectiflexibiles
PS-9	*Streptomyces violaceus*	White	+	+	+	Spiral
PS-14	*Streptomyces gibsonii*	White	–	–	–	Spiral

+: Presence; –: Absence.

Table 4.4: Utilization of Carbon Compounds by Three Strains of Antagonistic Actinobacteria

Strain No.	*Species to which Assigned*	*Utilization of Carbon Compounds*							
		A	*X*	*I*	*M*	*F*	*R*	*S*	*Ra*
PS-4	*Streptomyces moderatus*	+	+	+	+	+	+	+	±
PS-9	*Streptomyces violaceus*	–	–	+	+	+	+	+	+
PS-14	*Streptomyces gibsonii*	+	+	+	–	–	–	–	–

A: Arabinose; X: Xylose; I: Inositol; M: Manitol; F: Frutose; R: Rhamnose; S: Sucrose; Ra: Raffinose.

+: Presence; –: Absence; ±: Doubtful.

Conclusion

Present study indicates that the shrimp pond sediments are potential source for isolation of actinobacteria. The study also suggests clearly that the marine antagonistic actinobacteria are good candidate of produce antibiotics. Studies have been continued to isolate, purify and characterize the antibacterial compounds from the antagonistic marine actinobacteria and to test their possible use as alternative chemotherapeutic drugs and produce them on a commercial scale.

Acknowledgements

Authors thank Prof. T. Balasubramanian, Director, Centre of Advanced Study in Marine Biology, Annamalai University and the Head, Department of Biotechnology, Pondicherry University for providing with necessary facilities.

References

Balagurunathan, R. and Subramanian, A., 1998. *In vitro* inhibition of fish pathogens by an antibiotic from *Streptomyces griseobrunneus* (P–33). *Malays. Appl. Biol.*, 27: 149–150.

Balagurunathan, R., 1992. Antagonistic actinomycetes from Indian shallow sea sediments with reference to B-unsaturated α-lactone type of antibiotic from *Streptomyces griseobrunneus* (P–33). *Ph.D. Thesis*, Annamalai University, 82 pp.

Bauer, A.W., Sherris, J.C., Truck, M. and Kirby, W.H.M., 1966. Antibiotic susceptibility testing by standard single disc method. *Am. J. Clin. Path.*, 115: 493–496.

Buchanan, R.E. and Gibbons, V., 1974. *Bergey's Manual of Determinative Bacteriology*, 8th edn. The Williams and Wilkins Co., Baltimore, pp. 747–842.

Dalmin, G., Purushothaman, A. and Kathiresan, K., 2002. Distribution of total heterotrophic bacteria (THB) and *Vibrio parahaemolyticus* in shrimp culture ecosystem. *Indian J. Fish*, 49: 247–253.

Dhevendaran, K. and Annie, K., 1999. Antibiotic and L-asparaginase activity of streptomycetes isolated from fish, shellfish and sediments of Veli estuarine lake along Kerala coast. *Indian J. Mar. Sci.*, 28: 335–337.

Goodfellow, M. and Haynes, J.A., 1984. Actinomycetes in marine sediments. In: *Biological, Biochemical and Biomedical Aspects of Actinomycetes*, (Eds.) L. Ortiz-Ortiz, L.F. Bojalil and V. Yakoleff. Academic Press, Inc., Orlando, Fla, pp. 542.

Hameed, A.S.S., 1993. A study of the aerobic heterotrophic bacterial flora of hatchery reared eggs, larvae and post-larvae of *Penaeus indicus*. *Aquaculture*, 17: 195–204.

Janakiram, P., Jayasree, L. and Madhavi, R., 2000. Bacterial abundance in modified extensive and semi-intensive shrimp culture ponds of *Penaeus monodon*. *Indian J. Mar. Sci.*, 29: 319–323.

Kathiresan, K., Balagrunathan, R. and Masilamani, M., 2005. Fungicidal activity of marine actinomycetes against phytopathogenic fungi. *Indian J. Biotech.*, 4: 271–276.

Kurtboke, D.J. and Wildman, H.G., 1988. Accessing Australian biodiversity towards an improved detection of actinomycetes: An activity report. *Actinomycetes*, 9: 1–2.

Lechevalier, M.P. and Lechevalier, H., 1970. Chemical composition as a criterion in the classification of aerobic actinomycetes. *Int. J. System. Bacteriol.*, 20: 435–443.

Maclean, M.H., Ang, K.J., Brown, J.H., Jauncey, K. and Fry, J.C., 1994. Aquatic and benthic bacteria responses to feed and fertilizer application in trials with the freshwater prawn, *Macrobrachium rosenbergii*. *Aquaculture*, 120: 81–93.

Miyadoh, S., 1993. Research on antibiotic screening in Japan over the last decade: A producing microorganisms approach. *Actinomycetologica*, 9: 100–106.

Muthurayar, T., Sivakumar, K., Sahu, M.K., Thangaradjou, T. and Kannan, L., 2006. Mutational effect on the antibacterial activity of marine actinomycetes isolated from *Chanos chanos*, (Forskal, 1775). *Environ. Eco.*, 24: 46–50.

Nonomura, H., 1974. Key for classification and identification of 458 species of the *Streptomycetes* included in ISP. *J. Ferment. Technol.*, 52: 78–92.

Okami, Y., 1986. Marine microorganisms as a source of bioactive agents. *Microb. Ecol.*, 12: 65–78.

Otta, S.K., Karunasagar, I. and Karunasagar, V., 1999. Bacterial flora associated with shrimp culture ponds growing *Penaeus monodon* in India. *J. Aqua. Trop.*, 14: 309–318.

Patil, R., Jeyaskaran, G., Shanmugan, S.A. and Shakila, R.J., 2001. Control of bacterial pathogens, associated with fish diseases by antagonistic marine actinomycetes isolated from marine sediments. *Indian J. Mar. Sci.*, 30: 264–267.

Primavera, J.H., 1994. Environmental and socio economic effects of shrimp farming: The Philippine experience. *Infofish International*, 1/94: 44–49.

Sahu, M.K., Sivakumar, K. and Kannan, L., 2004. Estuarine fish as a source of antagonistic actinomycetes: An inventory. In: *Proceeding of the Conference on Microbiology of the Tropical Seas*, 13–15 December, National Institute of Oceanography, Goa.

Sahu, M.K., Sivakumar, K. and Kannan, L., 2005a. Isolation of actinomycetes from different samples of the Vellar estuary, Southeast coast of India. *Poll. Res.*, 24: 45–48.

Sahu, M.K., Sivakumar, K. and Kannan, L., 2005b. Degradation of organic matters by the extra-cellular enzymes of actinomycetes isolated from the sediments and molluscs of the Vellar estuary. *J. Aqua. Biol.*, 20: 142–144.

Sahu, M.K., Sivakumar, K. and Kannan, L., 2006. Isolation and characterization actinomycetes inhibitory to human pathogens. *Geobios*, 33: 105–109.

Sharmila, R., Abraham, T. Jawahar and Sundararaj, V., 1996. Bacterial flora of semi-intensive pond-reared *Penaeus indicus* (H. Milne Edwards) and the environment. *J. Aqua. Trop.*, 11: 193–203.

Shirling, E.B. and Gottlieb, D., 1966. Methods for characterization of *Streptomycetes* species. *Int. J. System. Bacteriol.*, 16: 313–340.

Sivakumar, K., Sahu, M.K. and Kathiresan, K., 2005a. Isolation and characterization of streptomycetes, producing antibiotic, from a mangrove environment. *Asian J. Microbiol. Biotech. Envi. Sci.*, 7: 87–94.

Sivakumar, K., Sahu, M.K. and Kathiresan, K., 2005b. Isolation of actinomycetes from the mangrove environment of the southeast coast of India. *Eco. Env. Cons.*, 11: 29–31.

Sivakumar, K., Sahu, M.K. and Kathiresan, K., 2005c. An antibiotic producing marine *Streptomyces* from the Pichavaram mangrove environment. *J. Annamalai Uni.*, 41(B): 9–18.

Sivakumar, K., Sahu, M.K., Manivel, P.R. and Kannan, L., 2006. Studies on L-glutaminase producing actinomycetes strain LG-10 from the estuarine fish, *Chanos chanos* (Forskal, 1775). *Indian J. Exp. Biol.*, 44: 256–258.

Takizawa, M., Colwell, P.R. and Hill, R.T., 1993. Isolation and diversity of actinomycetes in the Chesapeake Bay. *Appl. Environ. Microbiol.*, 59: 997–1002.

Umamaheswary, K., Sahu, M.K., Sivakumar, K., Thangaradjou, T., Sumitha, D. and Kannan, L., 2005. Investigations on L-glutaminase producing actinomycetes strain LG-33 from the estuarine fish, *Mugil cephalus* (Linnaeus, 1758). *Environ. Eco.*, 23: 942–947.

Waksman, S.A. and Lechevalier, H.A., 1962. *The Actinomycetes*. The Williams and Wilkins Co., Baltimore, U.S.A. Vol. III.

Chapter 5

The Dynamics of Gonad Growth and Ascorbate Status in Certain Commercially Valued Marine and Freshwater Fishes of Orissa

☆ *A.K. Patra*

Introduction

India is blessed with the vast and varied fish germplasm resources distributed widely in vivid aquatic ecosystem. In all oviparous fish, the developing embryo is totally dependent on nutrients stored in the egg yolk for successful development. An egg needs specific nutrients, such as ten indispensable amino acids, phospholipids (PLs) containing PUFA (Poly-unsaturated fatty acids), Vitamins, Calcium and trace elements to form a viable embryo (Mukhopadhyay *et al.*, 2003). These specialized nutrients must be available in sufficient quantities to the female brood stock during the growth of the oocyte in the ovary for subsequent incorporation in mature eggs to sustain development of viable embryo. During the maturation process, part of the dietary nutrients is oriented towards gonadal instead of somatic growth (Izquierdo *et al.*, 2000). In fact, various aspects of reproductive physiology in fish are intricately linked to gross nutrient availability and this regulating role of nutrition on reproduction have profound effects on fecundity, egg size, chemical composition of egg, hatchability and viability of larvae (Izquierdo *et al.*, 2000). Through manipulation of dietary factors, it is possible to improve reproductive performance as well as ensure steady supply of quality fish spawn throughout the year and not just during the annual monsoon spawning. In recent years, more systematic studies on the effect of different dietary constituents on brood stock performance and production of spawn are being carried out in important cultivable species around the world (Emata *et al.*, 2000; Shiau and Hsu, 2002; Sealey and Gatlin, 2002; Lee and Dabrowski, 2003).

Several studies on vitamin nutrition have demonstrated the importance of Vitamin A, E and C on several reproductive processes of fish (Izquierdo *et al.*, 2000, Lee and Dabrowski, 2003). Dietary ascorbic acid (AA) performs important functions in gonadal growth and maturation in fish. Ascorbic acid is involved in multi-step control of neurohormonal events in fish reproduction (Dabrowski and Ciereszko, 2001; Lee and Dabrowski, 2004). During gonadal development and maturation presence of high concentration of AA in gonads especially in the ovary suggests possible metabolic effects of AA in reproduction (Dabrowski and Lee, 2004). The increase of ovarian AA during ovarian growth followed by its decline in the last stages before ovulation implies possible demand of AA in the hydroxylating reactions for steroidogenesis in the ovarian follicle cells. In particular, AA acts as a co-factor as a regulator in the biosynthesis of oestrogen in the follicle cells. The level of circulating 17-cc-oestradiol increases with adequate dietary supply of AA, thus indicating a higher deposition rate of viterogenin (Lee and Darbrowski, 2004).

AA also acts as an antioxidant (Lee and Dabrowski, 2003). An increase in seminal plasma AA concentration reduces peroxidative damage to spermatozoa during the reproductive season which in turn affects fertilizing ability of sperm and more specifically the genetic integrity of gametes. During mitotic germ cell division irreversible damage occurs when the concentration of ascorbate in gonads is diminished. Ascorbate concentration in fish seminal plasma is directly affected by the dietary supply of Vitamin C and also by the season; ascorbate concentrations decline towards the end of the season (Dabrowski and Ciereszko, 2001). Ascorbate deficiency in fish significantly reduces the sperm concentration, total sperm production over the season and sperm mortality (Dabrowski *et al.*, 1995).

There appears to be a paucity of data on the aspects of role or involvement of ascorbic acid on the reproduction of some economically important marine and freshwater fish species of India viz; *Rastrelliger kanagurta, Pellona ditchefa. Etroplus surotensis, Anabas testudineus, Channa punctatus* and *Channa striatus*. In this context the aim of this work was to evaluate the status of ascorbic acid in different tissues of economically important freshwater and marine fish species in relation to their reproductive cycle and also to evaluate the ascorbic acid biosynthesizing capacity of these species by assaying 1-gulonolactone y-oxidase activity. This study might have great relevance in the brood stock management of the above said species for the enhancement of quality of gamete and embryo.

Materials and Methods

Collection of Both Marine and Freshwater Fish

Qualitative collection of commercially valued marine and important both marine and freshwater fish species (*Rastrelliger kanagurta Pellona ditchela Etroplus surotensis, Anabas testudineus, Clarius batrachus,* and *Heteropneusts fossilis)* were made during several extensive and intensive surveys spanned over several months during February to August. Most of the collections were obtained from fisherman operating in the area *viz.*, Konark (Lat. 19° 20' to 20°5' North; Lon. 86°2' to 86°15' East), Puri (Lat 19° 28' to 20° 10' North; lat 86°9' to 86°25') and Chilka (Lat. 19°28' to 19°54' North; Lon. 85°05′ Mo 85°38 East) and also from wholesale fish market at Bhubaneshwar (4 No. Market of Lat 20°15' North; Lon. 85°82' East). Fishes were collected in every 15 days during February to August to assess length, weight, Gonado Somatic Index (GSI), tissue ascorbic acid concentration, activity of 1-gulonolactone y-oxidase, alkaline phosphatase, acid phosphatase and DNA and RNA concentrations of different tissues. Fresh samples preserved in icebox were transported to the laboratory for the study of concentration of tissue ascorbic acid, activity of 1-gulonolactone y-oxidase, alkaline phosphatase, acid phosphatase and DNA and RNA content in different tissues.

For approximate estimate of catch composition, random samples from different landing centers were examined during different seasons, at the time of collection of samples for the study of the breeding biology. Random samples of the required species were collected roughly taking into consideration of all available size groups. The entire study period February-August was divided into three breeding seasons *viz.*, pre-breeding, breeding and post-breeding as these fish showed their peak spawning during April-July (breeding), February-March (pre-breeding) and August onwards (post-breeding) (Chondar, 1999).

Most of the collections for biological study came from bagnet fishery operations. Most of these measurements were carried out on fresh fishes immediately after collections and after measurements, different bio-chemical analysis of different organs were done in laboratory of Fishery and Aquaculture Unit, Department of Zoology, Utkal University. Species identification was based on Talwar and Kacker (1984).

Biological Study

Total length was measured from tip of snout to caudal fin and standard length from tip of snout to caudal peduncle. Weight was taken to nearest 500 mg. Gonad weight was taken. Gonado Somatic Index (GSI) is expressed in percent body weight.

$$GSI = \frac{(Gonad\ weight)}{Body\ weight} \times 100$$

Tissue Preparation for Biochemical Analysis

Both marine and freshwater fish from different regions after collection, were brought to the laboratory in icebox for different chemical analysis.

Analysis of Vitamins (Ascorbic Acid)

For ascorbic acid analysis liver, gonad, kidney and brain of collected marine and freshwater fishes were dissected out in cold room and analysed for ascorbic acid immediately. The fish and tissue were homogenised in 5 per cent trichloroacetic acid (TCA) solution containing 250 nM $HClO_4$ and 0.08 per cent ethylenediaminetetra acetic acid (EDTA) using an motor driven glass Teflon homoginiser in ice and centrifuged at 29,000 x g for 30 min at 4°C. Total ascorbic acid content in tissue homogenate were measured by 2,4 dinitro phenylhydrazine (DNPH) method by Roe and Kuether (1943) modified by Dabrowski *et al.* (1989) in which 2, 4 dinitrophenyl hydrazine derivative of ascorbate was measured spectrophotometrically at 524 nm. Modification of original method included incubation with dichloroindophenol (DCIP) shortened to 20 min, incubation temperature was lowered from 37 to 30°C and additional blank per sample was included to account for interfering substances. Analysis of 1-gulonolactone γ-oxidase:

For 1-gulonolactone γ-oxidase analysis liver and kidney of collected marine and freshwater fishes were dissected out in cold room and analysed for L-gulonolactone oxidase immediately. Activity of L-gulonolactone γ-oxidase was measured using the method as described by Mukhopadhyay *et al.* (2003) quantity of tissue was homogenized in a glass homogenizer with 2ml 50mM phosphate buffer (pH 7.4) containing 1mM pH EDTA and 0.2 per cent sodium deoxycholate. The homogenate was centrifuged for 30 minutes at 4°C at 5000 rpm to remove nuclei and cell debris. The supernatant was used as enzyme source. 50 µl of freshly prepared 100nM L-gulonolactone (Sigma Chemical Co. USA) in phosphate buffer was added to 2ml of supernatent. After incubation for 30 minutes at 30°C. the reaction was stopped by addition of 1ml ice-cold 0.25M perchloric acid. This was then centrifuged at 2000rpm for 10 minutes and the supernatent was assayed spectrophotometrically for ascorbic acid by above said method. Protein was assayed in supernatant by the method of Lowry *et al.* (1951).

Analysis of Alkaline Phosphatase

Alkaline phosphatase (EC.3.1.3.1) activity was quantitatively estimated following the method of Bramley (1974) with slight modification using p-nitrophenyl phosphate (P–NPP) in alkaline medium. The enzyme catalysed the hydrolysis of P-NPP in alkaline medium to form p-nitrophenol (P-NP) and inorganic phosphate. The intensity of the yellow colour formed due to the released P-NP (in the form of dissociated phenylate ion), proportional to the activity of alkaline phosphatase, was measured spectrophotometrically at 410 nm. The assay mixture consisted of 100 ul enzyme solution, 100 uj substrate, 600 µl buffer (0.1 M sodium carbonate pH 10.4), 100 ul EDTA (adjusted to pH 10.4 with 1N NaOH), 100 µl magnesium acetate, and incubated at 30°C for 30 minutes. The reaction was stopped by adding 3000 µl of 0.5 per cent (w/w) EDTA in 0.5 M NaOH after 30 minutes. The absorbance was measured at 410 nm and the quantity of P-NP produced measured using P-NP as standard. Protein content of supernatant solution was measured following the method of Lowry *et al.* 1951). Enzyme activity expressed as µg P-NP liberated/mg protein/hr. Analysis of Phosphatase was quantitavely made following the method of Venha-Puttuala and Nikka-nen (1973).

Water Quality Variables

The hydrobiological parameters of the water from different water bodies (collection area of fishes) such as pH, dissolved oxygen, ammonia, free carbondioxide, total alkalinity, total ammonia nitrogen were monitored according to standard procedures (APHA, 1989).

Extraction and Quantification of DNA and RNA

Tissues were dissected out from the fresh fishes and the nucleic acids (DNA and RNA) were extracted and quantified according to the methods followed by Mitra and Mukhopadhyay, (2002).

Results and Discussion

The length and weight of 6 species were measured. The length and weight of different species were presented in Table 5.1. Gonado Somatic index of ovary and testis of both marine and freshwater fishes increased during breeding season and decreased after breeding season and showed +ve correlation (r = 0.7) with ascorbic acid in the gonad. Ascorbic acid content was assayed in different tissues *viz.*, liver, kidney, gonad (testis and ovary) and brain in the pre-spawning, breeding and post-spawning seasons. Among the different tissues ascorbic acid content was highest in brain during breeding season in all the fishes. The ascorbic acid content in different tissues followed the trend as brain > kidney > liver > gonad through out the reproductive cycle. The brain ascorbic acid level showed no variation in the three seasons-Ascorbic acid concentration increased during breeding season and declined towards the end of the reproductive season.

Acid phosphatase activity in all gonad analysed was maximum in the breeding season in comparison to the other seasons. However, acid phosphatase activity was higher in female fishes than the male ones in all the freshwater and marine fish species. Again the marine fishes showed higher acid phosphatase activity in gonad in comparison with the freshwater species. Acid phosphatase activity in different fishes showed positive correlation (r = 8) with ascorbic acid in the diet. Alkaline phosphatase activity also followed the same trend as acid phosphatase in both the species. Alkaline

Table 5.1: Collection Site, Study Period and Length Weight of Different Fish Species (200)

Sl.No.	Name of Species		Collection Site	No. of Fish in Each Period		Study Period			Avg. Wt. (gm)	Avg. Length (cm)
	Common Name	Scientific Name		Male	Female	Pre-spawning	Spawning	Post spawning		
1.	Koi, Kau	*Anabas testudineus*	Village Pond (near Konark)	10	10	Feb. to April	April to June	August onwards	24 to 25	11 to 13
2.	Mam	*Clarias batrachus*	Village Pond (near Konark)	10	10	"	"	"	203 to 205	45 to 47
3.	Marua, Languda-Modi	*Rastrelliger kanagurla*	Coastal Area (Near Konark)	10	10	"	"	"	118 to 120	20 to 21
4.	Paniakhia	*Pellona ditchela*	Coastal Area (Near Konark)	10	10	"	"	"	100 to 101	20.5 to 21
5.	Cundahle	*Etroplus suratensis*	Chilka Lake	10	10	"	"	"	96 to 99	19 to 19

phosphatase activity in different fishes also showed positive correlation (r = 6) with ascorbic acid in the diet. Ratio of RNA, DNA in gonad also followed the trend as acid phosphatase and alkaline phosphatase in all the freshwater and marine fish species (both sexes). l-gulonolactone γ-oxidase activity showed the negative result (r = 3) in *Rastrelliger kanagurta, Pellona ditchela and Etroplus suratensis.* λ-gulonolactone γ-oxidase activity in other freshwater fish species was studied by Chatterje (1973).

Table 5.2: Water Quality Variables of Freshwater

pH	7.1–7.6
Dissolved Oxygen (mg/l)	4.11–5.01
Free CO_2 (mg/l)	3.0–4.2
Total Alkalinity (mg/l)	80.09–89.4
Total Ammonia Nitrogen (mg/l)	0.35–0.45

Table 5. 3: Water Quality Variables of Marine Water

pH	7.2–7.7
Salinity (ppt)	22.8–33.5
Dissolved Oxygen	4.5–5.1
Free CO_2 (mg/l)	3.1–3.9
Total Alkalinity (mg/l)	80.01–89.4
Total Ammonia Nitrogen (mg/l)	0.11–0.19

Ascorbic acid concentration in brain tissue of all fish species was the highest among different tissues during breeding cycle. The result of the study also similar with the result of Ciereszko and Dabrowski (1995). Ascorbic acid concentration declined, in other tissues towards the end of the reproductive season. The present findings in all fishes also corroborated with the results of Ciereszko and Dabrowski (1995). Among the molecules identified in teleost fish hypothalamus and hypophysis, nor adrenaline stimulates gonadotropin releasing hormone (GnRH) release from the hypothalamus. Whereas dopamine inhibits GnRH release from the hypothalamus as well as pituitary in vitro (Reeding and Patino, 1993). The release of gonadotropins (GtHs) (which results in gonad growth and ovulation) from the pituitary in vivo is inhibited by dopamine but stimulated by nor adrenaline. Ascorbate is considered a reductant in the dopamine β-hydroxylase reaction in the conversion of tyrosine to noradrenaline. Ascorbic acid may potentiate the inhibitory effects of dopamine on hypophysial hormone release (Skin and Stirling, 1988). High amount of ascorbic acid may accelerates–the synthesis of adrenaline in neurohypophysis which will result in enhanced secretion of GtH in the endocrine part. Towards the end of the reproductive season, testes regress and cell sensescence and apoptosis take place. This process may directly generate production of free oxygen. Lipid peroxidation of rich polyunsaturates in differentiating spermatocytes may also be responsible for production of oxygen radicals. It has been found that increased dietary ascorbic acid levels may protect against lipid peroxidation of rainbow trout spermatozoa (Liu *et al.*, 1997). High amount of ascorbic acid in diet might lower the amount malondialdehyde generated by sperm and increased the amount of docosahexaenoic acid in lipids. High concentrations of ascorbic acid in gonads of several fish species indicate a close relevance to reproduction (Dabrowski and Ciereszko, 2001). In yellow perch collected from the wild concentrations of total AA in liver and testis were significantly high (Dabrowski and

Ciereszko, 1996a, b). The increased fertilizing ability of sperm corresponded to the enhanced liver and testis AA concentrations. Even though the mechanism for AA's effect on sperm quality is not clear, studies with rainbow trout (Liu *et al.*, 1997) indicated that the mechanism could be based on the antioxidant function of AA. Several studies have demonstrated the antioxidant role of vitamin C in semen showing lower lipid peroxidation values in sperm membranes and lower malonaldehyde production in sperm in the presence of AA in fish and mammals. Also Fraga *et al.* (1991) demonstrated that human sperm DNA could be protected by Vitamin C, resulting in a lower value of oxidant degraded DNA excreted.

Acid phosphatase activity was increased during the breeding season and decreased after the breeding season in gonads of all fishes. The same result was found in rat (Males and Turkington, 1971; Venha-Perttula and Nikkanen, 1973). However, there is a paucity of data on the role of Acid phosphatase during breeding season offish. Acid phosphatase has been localized in the lysosomes of Sertoli cells and in the Golgi vacuoles and Cisternae of spermatogonia, spermatocytes and spermatids in the testes of rats (Tice and Barrnett, 1963) and mice (Dietert, 1966; Hugon and Borgers, 1966).

During the transformation of spermatids an intense acid phosphatase activity was observed in the luminal part of the sertoli cells, in which maturing spermatids were attached in rat (Chang *et al.*, 1974). At the time of spermatozoan release, the Sertoli cell might gave a strong reaction in lysosomes at the basal part of the cell. This enzyme has some activity during the later stages of acrosome formation in the Golgi apparatus (Chang *et al.*, 1974). Consistent with the observations of Posalaki *et al.* (1968) Chang *et al.* (1974) were also reported an increased acid phosphatase activity in multivesicular bodies in the cytoplasm of late maturing spermatids and adjacent anchoring Sertoli cells. In the Golgi apparatus of the latter cells, acid phosphatase was also found to be present. In many rodents, the observations of Dalcq (1967) suggested that acid phosphatase reaction was positive only in the acrosomal precursors, but in the mature spermatozoan this reaction disappeared. Acid phosphatase may be involved in the functional maturation of spermatozoa during the passage in the epididymis and finally in the vesiculation occurring between the outer acrosomal membrane and overlying sperm plasmamembrane (Lui and Meizel, 1979). During the sexual development of the rat, the level of acid phosphatase rapidly increased concurrently with the maturation of spermatogenesis (Males and Turkington, 1971; Vanha-Perttula and Nikkanen, 1973). Due to this close connection with the development of seminiferous tubules, acid phosphatase was suggested as a marker enzyme for the testis (Vanha-Perttula and Nikkanen, 1973). The increase in acid phosphatase activity in gonadal tissues in the present study might be similar with the above findings in rat and the activity is regulated by ascorbic acid concentration in the respective tissue.

In the present study higher alkaline phosphatase activity was obtained in all the tissues of both marine and freshwater fish during breeding season, than pre or post breeding season. Alkaline phosphatase, a brush border lysosomal enzyme is known to play a significant role in the transport of nutrient across cell membrane. This lysosomal hydrolytic enzyme system capable of removing inorganic phosphate from certain organic phosphate esters such as hexose phosphates (like glucose-6 phosphate), glycerophosphate and the nucleotides derived from the diet and nucleic acids by nucleases (Matusiewiz and Dabrowski, 1996). Matusiewiz and Dabrowski (1996) demonstrated the co-relation between the alkaline phosphatase activity and the ascorbic acid status in rainbow trout. Our results are also similar with the above findings that changes in alkaline phosphatase activity in different tissues has correlation with the tissue ascorbic acid content and the increased enzyme activity during breeding season was due to increased metabolic activity of different tissues.

Significant positive correlation was found between RNA/DNA ratio and tissue RNA content with ascorbic acid changes during breeding cycle. There are several studies in which tissue RNA concentration and RNA/DNA ratios have been used as indicators of recent growth or nutritional status in fish (Bulow *et al.*, 1978, Mitra and Mukhopadhyay 2002). Thus the RNA: DNA ratio indicated metabolic intensity which was deeply influenced by ascorbic acid content of the tissue. The importance for ascorbate in maintaining the integrity of DNA in ova and sperm might have great relevance (Dabrowski and Ciereszko, 2001). The importance of ascorbate in maintaining the integrity of DNA in ova and sperm may be relevant to fish as decreasing quality of eggs is frequently observed in older females (Kamler, 1992). Generation of oxidative stress in oocytes and surrounding ovarian cells relates to the aneuploidy (Dabrowski and Ciereszko, 2001).

Absence of 1-gulonolactone γ-oxidase activity confirmed the incapability of these species *viz.*, *Rastrelliger kanagurta, Pellona ditchela* and *Etroplus suratensis* to synthesize ascorbic acid (AA) de-novo. So these fish might depend on the exogenous supply.

Knowledge of tissue level of ascorbic acid content along with acid phosphatase, alkaline phosphatase, DNA and RNA level during breeding cycle of these different fish species might be important in assessing the involvement of the vitamin in overall metabolism during breeding. These data can be extrapolated in broodstock management of these economically important fish species for captive breeding and quality seed production.

Acknowledgement

The author is thankful to Head of the Dept. of Zoology, Utkal University for providing Laborotory facility and to Dr. Gopa Mitra and Pradynsini Mohapatra for their support.

References

APHA, 1989. *Standard Methods for the Examination of Water and Wastewater.* American Public Health Association, Washington, DC, 17.

Byrd, J.A., Pardue, S.L. and Hergis, B.M., 1993. Effect of ascorbate on luteinlzing hormone stimulated progesterone biosyntheses in chicken granulose cell *in vitro. Comparative Biochemistry and Physiology,* 104A: 279–281.

Bulow, J.A., Coburn, C.B. and Cobb, C.S., 1978. Comparison of two blue gill populations by means of the RNA-DNA ratios and liver-somatic index. *Transactions of American Fisheries Society,* 107: 799–803.

Chondar, S.L., 1999. *Biology of Finfish and Shellfish.* ScSc Publishers, West Bengal, India.

Chatterjee, I.B., 1973. Vitamin C synthesis in animals: Evolutionary trend. *Science and Culture,* 39: 210.

Ciereszko, A. and Dabrowski, K., 1995. Sperm quality and ascorbic acid concentration in rainbow trout semen are affected by dietary vitamin C: An across season study. *Biology of Reproduction,* 52: 982–988.

Chang, J.P., Yokoya'ma, M., Brinkley, B.R. and Mayahara, H., 1974. Electron microscopic cytochemical study of phosphatases during spermiogenesis in Chinese hamster. *Biol. Reprod.,* 11: 601–610.

Dabrowski, K. and Ciereszko, A., 2001. Ascorbic Acid and reproduction in fish: endocrine regulation and gamete quality. *Aquae. Res.,* 32: 1–19.

Dabrowski, K., Ciereszko, R.E., Blom, J.H. and Oltobra, J.S., 1995. Relationship between vitamin C and plasma concentrations of testosterone in female rainbow trout, Oncorhynchus mykiss. *Fish Physiology and Biochemistry*, 14: 409–414.

Dabrowski, K. and Ciereszko, A., 1996a. The dynamics of gonad growth and ascorbate status in yellow perch. *Aquaculture Research*, 27: 539–542.

Dabrowski, K. and Ciereszko, A., 1996b. Ascorbic acid protects against male infertility in a teleost fish. *Experientia*, 52: 97–100.

Dietert, S.E., 1966. Fine structure of the formation and fate of the residual bodies of mouse spermatozoa with evidence for the participation of lysosomes. *J. Morphol.*, 120: 317–346.

Dalce, A.M., 1967. Sur la cytochimie de l' idiosome et de l'acrosome chez les Rongue. C. R. Acad Sco D., 264: 2386–2391.

Emata, A.C., Borlongan, I.G. and Damaso, J.P., 2000. Dietary vitamin C and E supplementation and reproduction of milkfish *Chanos Chanos*. Forsskal. *Aqua. Res.*, 31: 557–564.

Fraga, C.G., Motchnik, P.A. Shigenaga, M.K., Helbock, H.J., Jacob, R.A. and Ames, B.N., 1991. Ascorbic acid protects against endogenous oxidative DNA damage in human sperm. In: *Proceedings of the National Academy of Sciences*, USA, 88: 11003–11006.

Hugon, J. and Borgers, M., 1966. Ultra structural and cytochemicai changes in spermatogonia and sertoli cells of whole-body irradicated mice. *Anat Rec.*, 155: 15–32.

Izquierdo, M.S., Fernandez-Palacios, H. and Tacon, A.G.J., 2001. Effect of broodstock nutrition on reproductive performance offish. *Aquaculture*, 197: 25–42.

Kamler, E., 1992. *Early Life History of Fish: An Energetics Approach*. Chapman and Hall, London, p. 267.

Lee, K.J. and Dabrowski, K., 2003. Interaction between Vitamin C and E affects growth, their tissue concentrations, lipid oxidation and deficiency symptoms in yellow perch, Perca flavescens. *Br. J. Nutr.*, 89: 589–596.

Lee, K.J. and Dabrowski, K., 2004. Long term effect and interactions of dietary Vitamins C and E growth and reproduction of yellow perch, *Perca flavescens*. *Aquaculture*, 230: 377–389.

Lowery, O.H., Resebrough, N.J., Farr, A.L. and Randall, R.J., 1951. Protein measurement with the Folin phenol reagent. *J. Biol. Chem.*, 193: 265–275.

Liu, L., Ciereszko, A. and Dabrowski, K., 1997. Dietary ascorbyl monophosphate depress lipid per oxidation in rainbow trout spermatozoa. *J. Aquatic Ani. Heal.*, 9: 247–257.

Lui, C.W. and Meizef, S., 1979. Further evidence in support of a role for hamster sperm hydrolytic enzymes in the acrosome reaction. *J. Exp. Zool.*, 207: 17–186.

Mukhopadhyay, P.K., Chattapadhyay, D.N. and Gopa, Mitra, 2003. Essentiality of broodstock nutrition to ensure quality fish seed production. *Infofish International*, 3: 25–31.

Mitra, G. and Mukhopadhyay, P.K., 2002. Growth nutrient utilization and tissue biochemical changes in Rohu, *Labeo rohita*, fed with natural and prepared diet. *J. Appl. Aquacult.*, 12: 65–80.

Males, J.L. and Turkington, R.W., 1971. Hormonal control of lysosomal enzymes during spermatogenesis in rat. *Endocrinology*. 88, 579–588.

Posalki, Z., Szabo, D., Bacsi, E. and Okros, I., 1968. Hydrolytic enzymes during spermatogenesis in rat: An electron microscopic and histochemical study. *J. Histochem. Cytochem.*, 16: 249–262.

Redding, J.M. and Patino, R., 1993. Reproductive physiology. In: *Physiology of Fishes*, (Ed.) Ditt Evans. CRC Press, Boca Raton Fl, p. 503.

Shiau, S.Y. and Hsu, C.Y., 2002. Vitamin E sparing effect by dietary Vitamin C in juvenile hybrid tilapia, *Oreochromis niloticus* X *O. aureus*. *Aquaculture*, 210: 335–342.

Sealey, W.M. and Gatlin, III. D.M., 2002. Dietary Vitamin C and E interact to influence growth and tissue composition of juvenile hybrid striped bass (*Morone chryspos* X *M. saxatilis*) but have limited effects on immune response. *J. Nutr.*, 132: 748–755.

Shin, S.H. and Stirling, R., 1988. Ascorbic acid potentiates the inhibitory effect of dopamine on porlactin release in primary cultured rat pituitary cells. *J. Endocrinology*, 18: 387–394.

Tice, L.W. and Barrnett, R.J., 1963. The fine structure localization of some testicular phosphatase. *Anal Rec.*, 147: 43–63.

Venha-Perttula, T. and Nikkanen, V., 1973. Acid phosphatases of the rat testis in experimental conditions. *Acta. Endocrinol.*, 72: 376–390.

Chapter 6

Effect of Feeding Dietary Protein Sources on Daily Excretion in *Channa punctatus* for Sustainable Aquaculture

☆ *Meenakshi Jindal, S.K. Garg and N.K. Yadava*

ABSTRACT

To study the daily excretion patterns of wastes like ammonical nitrogen (NH_4-N) and ortho-phosphate (O-PO_4) in fish *Channa punctatus*, the fingerlings were fed on 10 iso-nitrogenous diets (1 to 10) formulated by replacing fishmeal (FM) with defatted canola and processed soybean at 2 inclusion levels 75 and 100g/kg with and without supplementing the diets with a mineral premix and amino acids (MPA). Studies have revealed that oxygen levels (DO) fluctuated between 4 to 5 mg/l and pH remained alkaline (7.4 to 7.8). Significantly ($p<0.05$) highest conc. of NH_4-N and O-PO_4 were observed in the water medium in which fish were fed on reference diets 1 and 6 containing FM as the main protein source. The excretion decreased on increasing the inclusion levels of processed soybean and defatted canola respectively. The groups of fish fed on diets containing processed soybean as the main protein source shows lower excretion of NH_4-N and O-PO_4 in comparison to groups of fish fed on diets containing defatted canola as the main protein source. Further, in the groups of fish fed on diets 6 to 10 the excretion of NH_4-N and O-PO_4 was lower than those observed in groups of fish where diets 1 to 5 were used indicating that incorporation of MPA reduces the excretion of NH_4-N and O-PO_4 in water medium. Thus, excretion of nitrogen and phosphorous can be reduced by the use of protein of plant origin in feed which has important implications on the management of highly intensive farming system

Introduction

The aim of aquaculture should be to provide sufficient nitrogen for good growth through balanced feed. Nitrogen pollution from aquaculture can occur in 3 ways, *viz.* (1) overfeeding of fish at a time when they are not growing (2) feeding unstable and highly soluble diets, (3) providing a diet of poor absorption and nitrogen retention efficiency. Usually, protein is the most expensive component of aquaculture diets.

Fishmeal (FM), which is considered as the best source of protein for fish is difficult to get in interior parts and even if it is available, it will be very costly. Therefore, many plant proteins such as soybean, canola, guar, moong, sorghum etc. offer considerable promise in this regard owing to their low prices and market availability. Due to world wide dominance of soybean (*Glycine max*) and its appreciation as quality protein, several workers have attempted to replace FM with soybean meal in diets formulated for several fish species (Jindal and Garg, 2005; Robinson and Menghe, 2007).

Canola (*Brassica napus)* is primarily used for edible purposes, while the defatted meal is utilized as animal feed. It has about 40 per cent protein (on a dry matter, oil free basis) and a relatively well balanced amino-acid composition. It is a high protein feed ingredient of plant origin, however, it possesses some anti-nutritional compounds such as glucosides and tannins. Furthermore, protein in canola is surrounded by relatively indigestible carbohydrate, which cannot be broken down without the use of added enzymes (Buchman *et al.*, 1997).

The main product excreted by teleost fish is total ammonia nitrogen (TAN), which is formed in the lever and excreted across the gills. About 80-90 per cent of nitrogen loss from fish is through gill excretion and the faecal nitrogen loss accounts for 10-20 per cent. Nitrogen is also lost through uneaten feed or dust (Kibria *et al.*, 1998). Fish excrete phosphorus in soluble and particulate forms (Vielman *et al.*, 2000.) The soluble fraction is called ortho phosphate (O-PO_4), is most available for plant growth. However the main loading of phosphorus to the environment was reported to be via faecal pellets (Kibria *et al.*, 1998) (Figure 6.1).

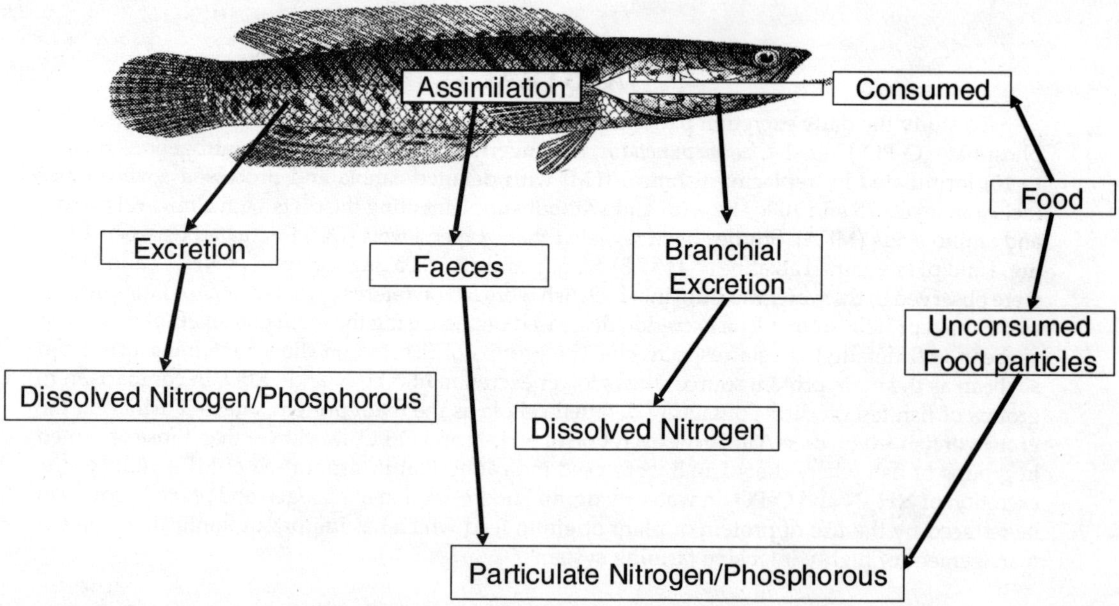

Figure 6.1: Nitrogen and Phosphorous Excretion/Pollution by *C. punctatus* in Aquaculture (Kibria *et al.*, 1996, 1998)

Various attempts have been made to search alternate protein sources, that may also reduces the input of nitrogen and phosphorus into the environment. Hence, the aim of present investigation was to search an alternate protein source of plant origin *i.e.* soybean that will not be only cost effective, but certainly would reduce excretion of nitrogenous wastes and total organic matter, possibly also of phosphorus, and alleviate pollution problems in intensive aqua-cultural systems.

Materials and Methods

Fish Species

Specimens of *Channa punctatus* were obtained from fish dealers of Hisar. Specimens with mean body weight (8.0 to 15.0 g) were used in the studies. Fish were placed in the transparent glass aquaria (60×30×30 cm) kept in the laboratory where the temperature was maintained at 25±1°C and the lighting scheduled at 12h of light alternating with 12h of darkness. The fish were acclimatized for a minimum of 7 days prior to the initiation of experimental treatments. The water was renewed daily with chlorine free water.

Preparation of Diets

Soybean and canola were used as the main protein sources. Soybean seeds were cleaned, autoclaved for half an hour at 121.6°C at 15 lb pressure to remove antinutritional factors (ANFs) such as trypsin inhibitors, haemoglutinins, lectins and phytic acid. (Garg *et al.*, 2002). After oven drying at 60°C, it was ground into fine powder. Canola was also defattened to remove antinutritional compounds prior to the preparation of experimental diets (Garg *et al.*, 2002).

Groundnut oil cake, rice bran, fishmeal (FM), defatted canola (DFC) and processed soybean (HPS) were finely ground to pass through 0.5 mm sieve. All the ingredients were mixed according to Table 6.1 and dough was made using distilled water. Thereafter, the dough was passed to a mechanical palletizer to obtain pellets (0.5 mm thick) which were dried in an oven and used in the studies for 45 days.

The experimental diets were fed to duplicate groups of fish to satiation twice a day for a one week acclimatization period before starting the study. After this period, the fish were individually weighed and their initial weights recorded. The fish were then offered the test diets (1-10) twice a day (9:00h and 16:00h) to satiation, for 45 days. This period was considered enough to produce the effect of feeding on daily excretory pattern in the test species. Faeces were siphoned from culture aquaria every morning before fish feeding. In addition, about 20-50 per cent of the culture water was replaced daily with new, fresh, dechlorinated water. Growth parameters like live weight gain, growth per cent gain, feed conversion ratio (FCR), apparent protein digestibility (APD) and protein efficiency ratio (PER) were calculated using standard methods given by Steffens (1989).

Water Analysis

The various water quality parameters like dissolved oxygen, pH, temperature, conductivity, free carbon dioxide, total alkalinity and total hardness of aquaria water were analyzed following APHA (1998).

Determination of Ammonical Nitrogen (NH_4-N) and Ortho-phosphate (O-PO_4)

On the last day of experiment offer the same feed to the fish in sufficient quantity so that the same is consumed, wait for 2 hours. Maintain a fixed level of water in each aquarium (say 30–40 L). Remove the excess of feed. Start collecting water samples from each aquarium in replicate of two for the determination of ammonical nitrogen (NH_4-N) and ortho-phosphate (O-PO_4) following APHA (1998) to see the influence of compounded feeds on pollution status of receiving water in the aquaria.

Table 6.1: Ingredient and Proximate Composition (Per cent dry weight basis) of Experimental Diets (1 to 10) Containing Fishmeal, Processed Soybean and Processed Canola as the Main Protein Sources

Ingredients	*Diet Number*									
	1	*2*	*3*	*4*	*5*	*6*	*7*	*8*	*9*	*10*
Groundnut Oil Cake	60.0	60.0	60.0	60.0	60.0	60.0	60.0	60.0	60.0	60.0
Rice Bran	24.0	24.0	24.0	24.0	24.0	23.0	23.0	23.0	23.0	23.0
Fish Meal (FM)	10.0	2.5	–	2.5	–	10.0	2.5	–	2.5	–
Hydrothermically processed Soybean (HPS)	–	7.5	10.0	–	–	–	7.5	10.0	–	–
Processed Canola (DFC)	–	–	–	7.5	10.0	–	–	–	7.5	10.0
Chromic Oxide (Cr_2O_3)*	1.0	1.0	1.0	1.0	1.0	1.0	1.0	1.0	1.0	1.0
Binder (Carboxyl methyl cellulose)	5.0	5.0	5.0	5.0	5.0	5.0	5.0	5.0	5.0	5.0
MPA**	–	–	–	–	–	1.0	1.0	1.0	1.0	1.0

*: Cr_2O_3 was used for estimating apparent digestibility.

**: mineral premix and amino acid (MPA) supplemented @ 10gkg^{-1} diet.

Each Kg contains Cu: 312 mg; Co: 45 mg; Mg: 2.114 g; Fe: 979 mg; Zn: 2.13 g; I: 156 mg; DL-Methionine: 1.92g; L-lysine mono hydrochloride: 4.4 g; Ca: 30 per cent; P: 8.25 per cent.

Diets: (1) 100 per cent FM; (2) 75 per cent HPS; (3) 100 per cent HPS; (4) 75 per cent DFC; (5) 100 per cent DFC; (6) 100 per cent FM + MPA; (7) 75 per cent HPS + MPA; (8) 100 per cent HPS + MPA; (9) 75 per cent DFC + MPA; (10) 100 per cent DFC +MPA.

Calculate the excretory levels of NH_4-N and O-PO_4 in treated water as follows:

$$\text{NH}_4\text{-N excretion (mg/100g BW of fish)} = \frac{\text{NH}_4\text{-N (mg l}^{-1}\text{) in aquarium water}}{\text{Fish weight (mg) per L of water}}$$

$$\text{O-PO}_4\text{ excretion (mg/100g BW of fish)} = \frac{\text{O-PO}_4\text{ (mg l}^{-1}\text{) in aquarium water}}{\text{Fish weight (mg) per L of water}}$$

Statistical Analysis

Data was analysed following ANOVA followed by Duncan multiple range test (Duncan, 1955) at 5 per cent probability level. Group means were compared by student 't' test.

Results and Discussion

Water Quality Parameters

The monitoring of various physico-chemical parameters of water used throughout the experimental period revealed that level of dissolved oxygen was in the range of 4.2 to 4.9 mg/l, pH remained alkaline *i.e.* around 7.2 to 7.6, conductivity was in the range of 0.47 to 0.52 micro hos/cm, free carbondioxide was between 16.4 to 17.2 mg/l, alkalinity was between 219 to 253 mg/l, hardness was between 211 to 233 mg/l (Table 6.2). The ambient temperature in the laboratory was maintained between 25 to 27°C. These conditions were quite optimum for normal fish activity and growth.

Post prandial excretory levels of ammonaical nitrogen (NH_4-N) and ortho-phosphate (O-PO_4)

Significantly ($p< 0.05$) highest conc. of NH_4-N and O-PO_4 were observed in the water medium in which fish were fed on reference diets (1 and 6) containing FM as the main protein sources (Table 6.3).

Further in the groups of fish fed on diets 1 to 5, the excretion of NH_4-N and O-PO_4 was higher than those observed in groups of fish where diets 6 to 10 were used indicating that incorporation of MPA reduces the excretion of NH_4-N and O-PO_4 in water medium (Table 6.3). These results are in agreement with those of Viola and Lahav (1993). According to them the calculated amounts of excreted (not retained) nitrogen per kg gain was reduced by 20 per cent in the lysine supplemented feeds, as compared to the 30 per cent protein feed. Concomitantly, calculated phosphorus excretion per kg gain was also decreased approximately by 100 per cent.

The excretion of NH_4-N and O-PO_4 was Significantly ($p< 0.05$) lower in those groups of fish where processed soybean was used as the main protein source *i.e.* diets 2, 3, 7, and 8 in comparison to those groups of fish where defatted canola (diet 4, 5, 9 and 10) and FM (diet 1 and 6) were used as the main protein source (Table 6.3) respectively. This shows that fish fed on diets containing defatted canola (with and without supplementation of MPA), excreted higher amounts of NH_4-N and O-PO_4 in receiving waters, which may be attributed to slightly high protein contents in the canola based diets. These results are in agreement with those of Kalla *et al.* (2003), Jindal and Garg (2005), Kalla and Garg (2004), Jindal *et al.* (2007a,b), Jindal (2008). These results showed that processed soybean is the most promising plant based protein source and can be used in the diets of fish.

The excretion decreased on increasing the inclusion levels of defatted canola and processed soybean. This is because the fish can digest plant proteins much easier than animal proteins (Priyanka and Garg 2002 and Kalla and Garg 2004).

Diurnal Pattern of NH_4-N and O-PO_4 Excretion

Water samples were analysed for 16 hrs at 2 hr interval of post-feeding revealed peaks in NH_4-N and O-PO_4 excretion (Figure 6.2 to 6.5).

The peak time of excretion of NH_4-N in groups of fish fed on diets 1-5 were observed at the end of 8 hrs. of post feeding (Figure 6.2). In the groups of fish fed on diets 6 to 10 supplemented with MPA, the excretion of NH_4-N was lower than those observed in groups of fish fed on diets 1–5. But the pattern of excretion of NH_4-N was same as for groups of fish fed on diets 1-5 (Figure 6.3).

The peak time of excretion of O-PO_4 in the groups of fish fed on the diets 1 to 5 were observed at the end of 6 hrs. of post-feeding (Figure 4), whereas in the groups of fish fed on diets 6 to 10 supplemented with MPA, the excretion of O-PO_4 was lower than those observed in the groups of fish fed on diets 1 to 5. But the pattern of excretion of O-PO_4 was same as for the groups of fish fed on diets 1 to 5 (Figure 6.5).

These results are in agreement with those of Kaushik and Gomes, 1998; Jindal, 2007 and 2008, who reported ammonia excretion peaks between 7-9 hr post feeding. A peak in NH_4-N and O-PO_4

Table 6.2: Water Quality Parameters of Different Aquariums Stocked with *C. punctatus* Fingerlings Fed on Diets (1–10) Containing Fishmeal, Processed Soybean and Defatted Canola as the Main Protein Sources

Parameters	Diet No.									
	1	2	3	4	5	6	7	8	9	10
Dissolved oxygen (DO) mg/l	4.2± 0.003	4.4± 0.001	4.4± 0.002	4.6± 0.001	4.8± 0.000	4.4± 0.002	4.0± 0.001	4.3± 0.003	4.2± 0.000	4.9± 0.001
pH	7.2	7.2	7.4	7.5	7.5	7.4	7.3	7.3	7.6	7.5
Water temperature (°C)	25.0	26.0	25.0	25.0	27.0	26.0	25.5	26.5	26.0	25.0
Conductivity micro (μ) mhos cm^{-1}	0.513± 0.001	0.499± 0.000	0.521± 0.002	0.473± 0.000	0.489± 0.000	0.493± 0.003	0.511± 0.004	0.507± 0.002	0.503± 0.001	0.497± 0.000
Free carbondioxide (Free CO_2) mg/l	17.20± 0.003	17.03± 0.001	16.43± 0.003	16.86± 0.001	17.13± 0.000	16.40± 0.003	17.00± 0.001	16.89± 0.001	17.09± 0.002	16.92± 0.001
Total alkalinity (mg/l)	247.36± 0.003	249.37± 0.002	226.71± 0.004	237.60± 0.003	250.88± 0.004	219.30± 0.000	252.73± 0.003	222.67± 0.001	240.71± 0.000	252.67± 0.003
Total hardness (mg/l)	224.00± 0.004	212.81± 0.002	232.69± 0.000	219.73± 0.003	229.96± 0.001	211.81± 0.003	215.31± 0.002	219.01± 0.004	230.81± 0.001	222.63± 0.003

All values are mean±S.E. of means of 3 observations.

Table 6.3: Growth Performance of Fish *Channa punctatus* Fed on Ten Experimental Diets (1–10) Containing Fish Meal, Processed Soybean and Processed Canola Meal as the Main Protein Sources

Diet No.	*Weight Gain (g)*	*Growth % Gain*	*FCR*	*APD*	*PER*	*NH_4-N (mg g^{-1} BW of fish)*	*O-PO_4 (mg g^{-1} BW of fish)*
1	4.473^{A}±0.049	80.052^{B}±0.299	2.868^{F}±0.032	75.953^{A}±0.339	0.142^{A}±0.001	0.606^{H}±0.003	0.276^{G}±0.006
2	5.223^{F}±0.033	86.398^{F}±0.322	2.518^{D}±0.016	82.143^{F}±0.380	0.173^{C}±0.001	0.496^{D}±0.003	0.226^{E}±0.008
3	5.360^{G}±0.105	87.197^{G}±0.717	2.476^{C}±0.047	85.297^{H}±0.104	0.182^{D}±0.005	0.473^{C}±0.003	0.190^{D}±0.005
4	4.852^{C}±0.043	81.856^{C}±0.327	2.700^{E}±0.024	79.129^{D}±0.426	0.150^{A}±0.001	0.516^{E}±0.000	0.186^{D}±0.003
5	5.020^{D}±0.006	84.689^{D}±0.270	2.591^{D}±0.003	78.450^{C}±0.264	0.162^{B}±0.000	0.515^{E}±0.008	0.160^{B}±0.005
6	4.583^{B}±0.171	79.839^{A}±1.083	2.844^{F}±0.103	76.827^{B}±0.088	0.144^{A}±0.004	0.540^{G}±0.003	0.240^{F}±0.000
7	5.676^{H}±0.017	90.797^{H}±1.314	2.381^{B}±0.075	83.517^{G}±0.280	0.187^{E}±0.007	0.453^{B}±0.003	0.190^{D}±0.000
8	5.983^{I}±0.113	93.883^{I}±0.815	2.282^{A}±0.042	87.990^{I}±0.324	0.199^{F}±0.005	0.426^{A}±0.008	0.173^{C}±0.003
9	5.193^{E}±0.013	85.978^{E}±0.089	2.549^{D}±0.006	80.819^{E}±0.160	0.167^{B}±0.000	0.523^{EF}±0.003	0.163^{B}±0.003
10	5.669^{H}±0.015	90.788^{H}±1.312	2.379^{B}±0.074	83.515^{G}±0.278	0.185^{D}±0.006	0.480^{C}±0.000	0.110^{A}±0.005

All values are mean±S.E. of mean.

Mean with same letter in the same column are not significantly ($p>0.05$) different.

Data were analysed by Duncan's Multiple Range Test.

excretion at the end of 6 hr of post feeding has also been reported by Priyanka and Garg 2002; Kalla *et al.*, 2003; Singh *et al.*, 2003; Kalla and Garg 2004; Jindal and Garg, 2005; Jindal *et al.* (2007a,b).

Growth and Digestibility

Feeding results have revealed low mortality in all dietary treatments. Results have demonstrated the suitability of HPS and DFC as an alternate protein source for the fish *C. punctatus*. Studies have clearly revealed that FM can be partially (up to 75 per cent) replaced with the use of HPS and DFC in the diets of *C. punctatus*. Studies have further demonstrated that complete replacement of FM was possible when diets were supplemented with MPA and there was no deleterious effect on the growth of the fish. These results are in agreement with many recent studies where FM has been totally replaced with the use of soybean meal supplemented with MPA (Garg *et al.*, 2002; Kalla and Garg, 2004). Similarly, canola has also been successfully used for the total replacement of FM (Priyanka and Garg, 2002; Kalla and Garg, 2004; Jindal and Garg, 2005). Though no significant ($p>0.05$) differences in growth and digestibility parameters were observed in fingerlings fed on soybean and canola diets, however, live weight gain was significantly ($p<0.05$) high in fish fed on soybean based diets (Table 6.3).

The apparent protein digestibility (APD) and energy retention increased with increase in inclusion levels of plant proteins in the diets. These results are similar to those observed on a cyprinid *Labeo rohita* and *C. mrigala* (Kalla and Garg, 2004) and *C. batrachus* (Jindal, 2008). Studies have further revealed that APD had no effect on protein and energy retention of the fish. In general, the pattern of APD corresponds to growth trends of fish fed on different diets incorporating plant origin proteins.

Present studies have revealed that growth and digestibility in *C. punctatus* was further enhanced when DFC and HPS protein based diets were supplemented with MPA and significantly ($p<0.05$) higher growth was observed in fish fed on a diet where FM was completely replaced with HPS and DFC (diet 8 and 10). These results are in agreement to those of Kalla and Garg, 2004; Priyanka and Garg, 2002; Jindal and Garg, 2005), who have succeeded in achieving complete replacement of FM by the use of HPS and DFC.

Protein efficiency ratio (PER) values were observed significantly ($p<0.05$) high in fish fed on diets containing HPS and/DFC in comparison to the fish fed on FM containing diets.

However, when the diets were supplemented with MPA, a significantly ($p<0.05$) high live weight gain, low feed conversion ratio (FCR) and high PER and APD values were observed in fishes fed on HPS as compared to the fish fed on DFC based diets.

The studies further showed that growth and digestibility parameters were found to be negatively correlated with NH_4–N and O-PO_4 excretion (Table 6.3). This is also the reason for the better growth of fish fed on diet 8 and 10.

Comparison of FM, DFC and HPS Containing Diets

Studies have revealed high growth performance, nutrient retention, low FCR and low post prandial excretory levels of wastes (NH_4-N and O-PO_4) in the groups of fish fed on diets containing HPS as the protein source, which is followed by another group of fish fed on diets DFC as the main protein source, which is again followed by groups of fish fed on FM containing diets.

These studies thus clearly indicate that soybean appears to be more suitable protein source in comparison to DFC or FM.

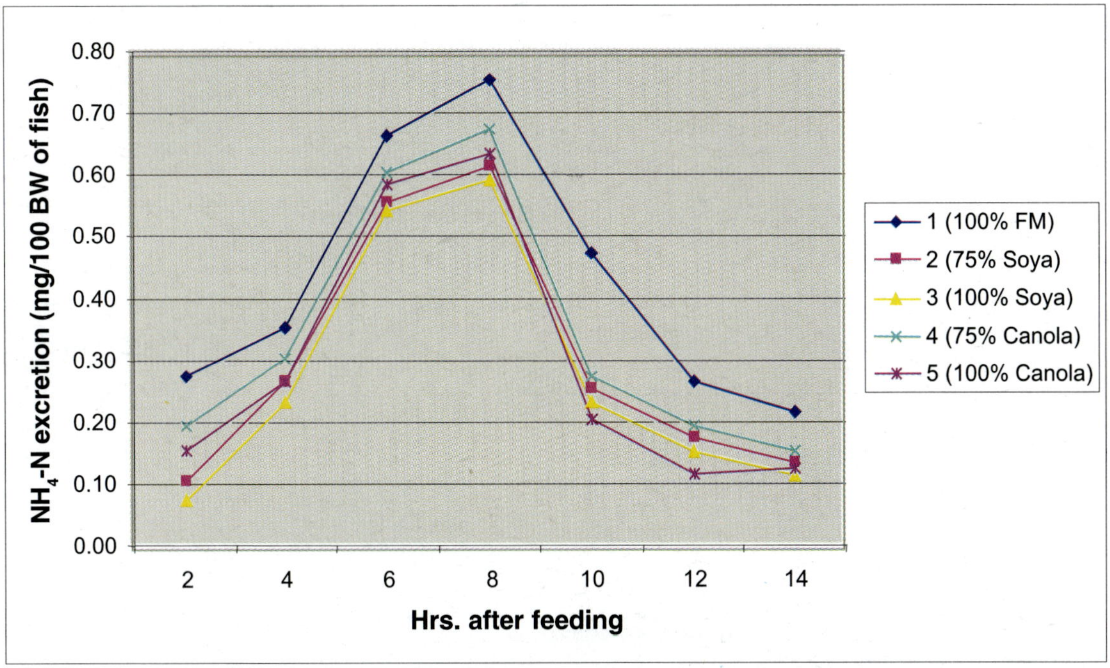

Figure 6.2: Excretion Pattern of Ammonical Nitrogen (NH_4-N; mg/100g body weight of fish) in Treated Waters in Fish *C. punctatus* Fed on Diets (1–5) without MPA

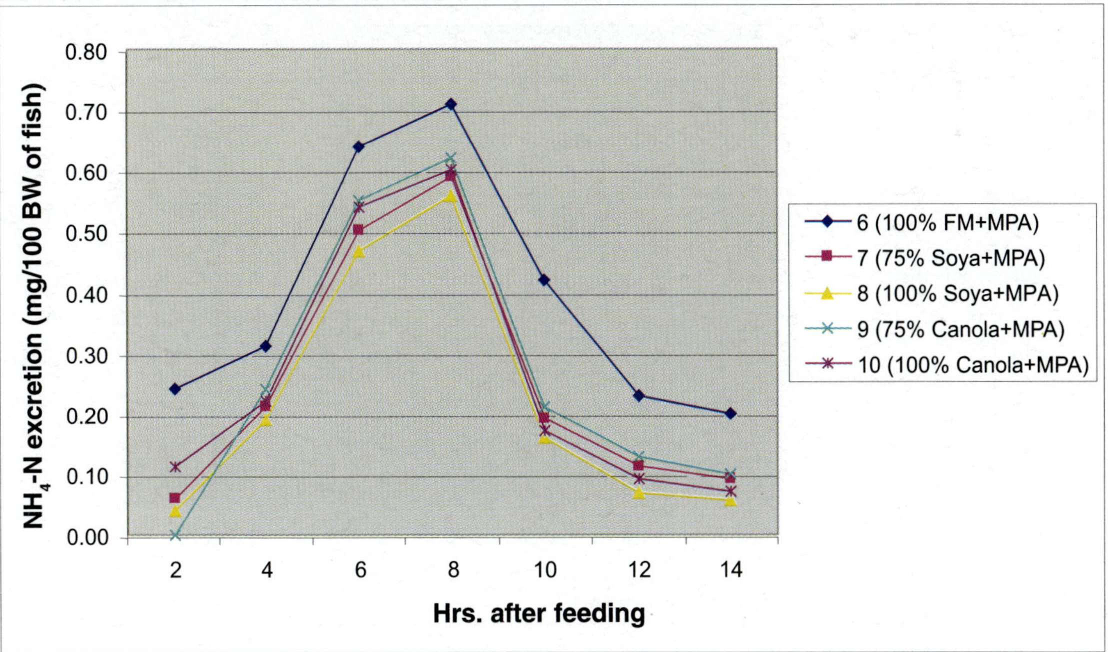

Figure 6.3: Excretion Pattern of Ammonical Nitrogen (NH_4-N; mg/100g body weight of fish) in Treated Waters in Fish *C. punctatus* Fed on Diets (6–10) with MPA

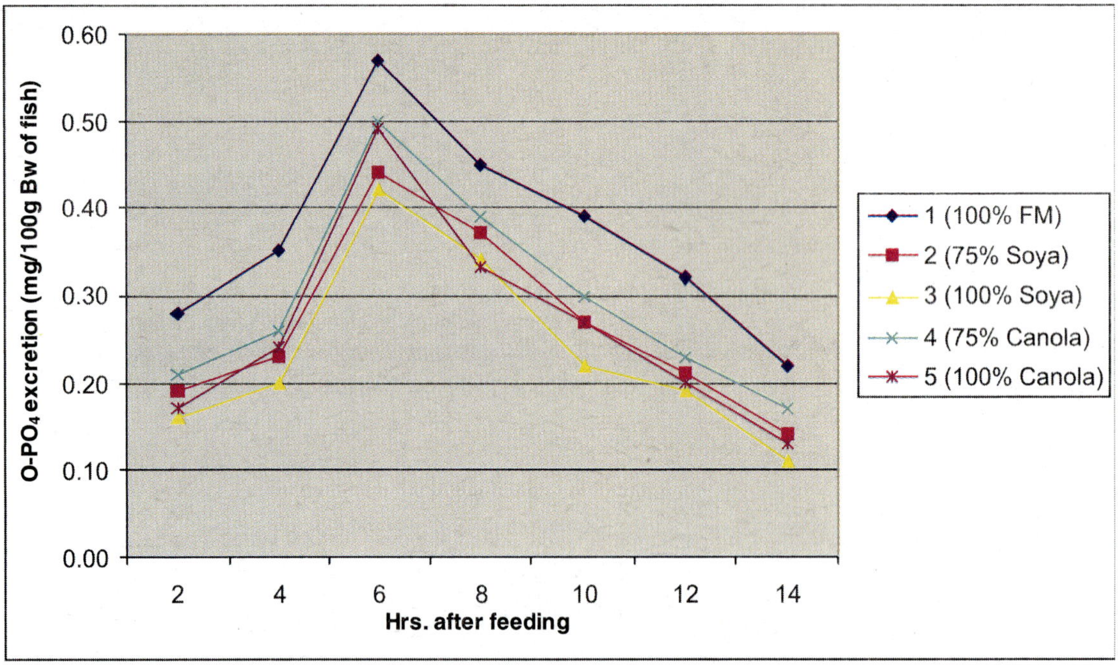

Figure 6.4: Excretion Pattern of Ortho-phosphate (O-PO_4 mg/100g body weight of fish) in Treated Waters in Fish *C punctatus* Fed on Diets (1 5) without MPA

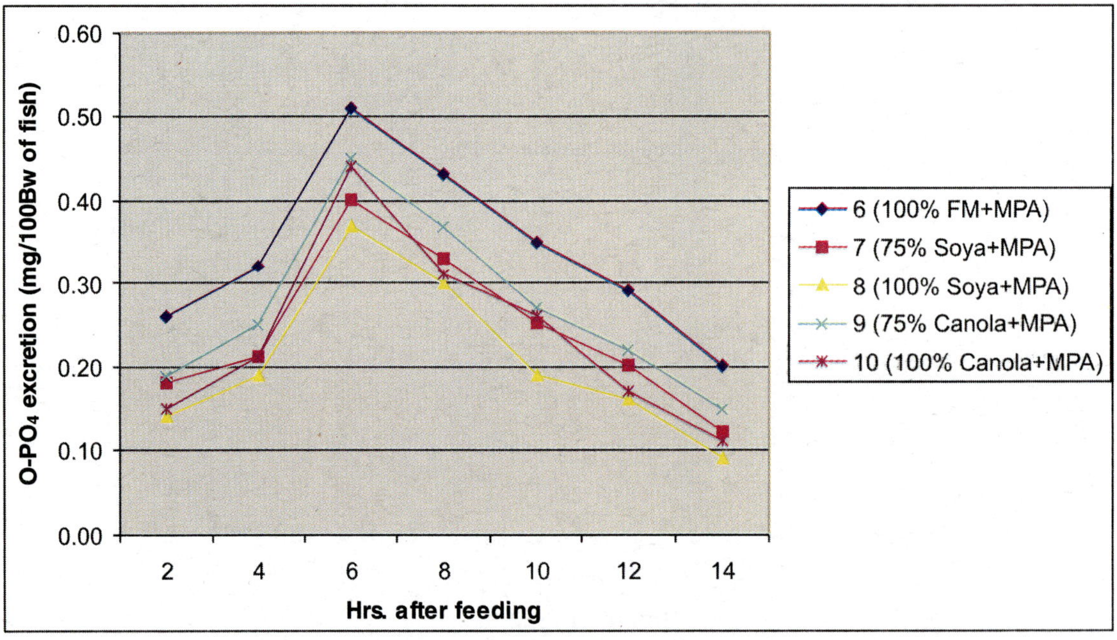

Figure 6.5: Excretion Pattern of Ortho-phosphate (O-PO_4 mg/100g body weight of fish) in Treated Waters in Fish *C. punctatus* Fed on Diets (6–10) with MPA

Conclusion

Protein is typically the most costly nutrient in a formulated feed. Protein in the majority of formulated fish diets (worldwide) depends greatly on FM, which is more costly than high quality, plant based protein sources, such as soybean protein. Nitrogen in waste water from aquaculture effluents is often considered a pollutant. In freshwater systems, nitrogen is sometimes a limiting nutrient, so adding it stimulates plant and algal growth. A majority of the excess nitrogen in either tank or pond culture systems originates as ammonia excreted by fish. The ammonia, as a waste product, is formed during the breakdown of proteins and excess amino-acids not incorporated into tissue by the fish.

HPS and DFC can be used as the main protein source of plant origin in the diets of *C. punctatus* for enhancing growth, digestibility and nutrient retention. But soybean appears to be more suitable protein source than canola. Such a replacement could save not only total feed costs but certainly would reduce excretion of nitrogenous and total organic matter, possibly also of phosphorous and alleviate pollution problems in intensive aquaculture systems.

Acknowledgement

Chaudhary Charan Singh Haryana Agricultural University is thankfully acknowledged for providing necessary facilities.

References

APHA (American Public Health Association), 1998. *Standard Methods for the Examination of Water and Wastewater*, 20th Edn., APHA, AWWA, EPFC, New York.

Buchman, J., Sarac, H.Z., Poppi, D. and Cowman, R.T., 1997. Effects of enzyme addition to canola meal in prawn diets. *Aquaculture*, 151: 29–35.

Duncan, D.B., 1955. Multiple and multiple F-tests. *Biometrics*, 11: 1–42.

Garg, S.K., Kalla, Alok and Bhatnagar, Anita, 2002. Evaluation of raw and hydrothermically processed leguminous seeds as supplementary feed for the growth of two Indian Major carp species. *Aquaculture Res.*, 33: 151–163.

Jindal, M., Garg, S.K.and Yadava, N.K., 2007b. Effect of replacement of fishmeal with dietary protein sources of plant origin on the growth performance and nutrient retention in the fingerlings of *Channa punctatus* (Bloch.) for sustainable aquaculture. *Pb. Univ. Res. J. (Sci.)*, 57 (in press).

Jindal, M, Garg, S.K., Yadava, N.K. and Gupta, R.K., 2007a. Effect of replacement of fishmeal with processed soybean on growth performance and nutrient retention in *Channa punctatus* (Bloch.) fingerlings. Livestock Research for Rural Development. Volume 19, Article #165. Retrieved from http://www.cipav.org.co/lrrd/lrrd19/11/jind19165.htm.

Jindal, Meenakshi and Garg, S.K., 2005. Effect of replacement of fishmeal with defatted canola on growth performance and nutrient retention in the fingerlings of *Channa punctatus (Bloch.) Pb. Univ. Res. J. (Sci.)*, 55: 183–189.

Jindal, Meenakshi, 2008. Studies on protein requirements of catfish *Clarias batrachus* fingerlings for sustainable aquaculture. In: *National Seminar on Technical Advances in Environment Management and Applied Zoology*, Jan. 23–25, Department of Zoology, Kurukshetra University, Kurukshetra, pp. 59–62.

Kalla, Alok and Garg, S.K., 2004. Use of plant proteins in supplementary diets for sustainable aquaculture. In: *National Workshop on Rational Use of Water Resources for Aquaculture*, (Eds.) S.K. Garg and K.L. Jain. March 18–19, Hisar, India, pp. 31–47.

Kalla, Alok, Garg, S.K. and Kaushik, C.P., 2003. Effect of dietary protein source on growth, digestibility and body composition in the fingerlings of *Cirrhinus* mrigala (Ham.) In: *Proceedings of 3rd Interaction Workshop on Fish Production Using Brackishwater in Arid Ecosystem*, (Eds.) S.K. Garg and A.R.T. Arasu. December 17–18, Hisar, India, pp. 139–145.

Kibria, G., Nugegoda, D., Fairclough, R. and Lam, P., 1998. Can nitrogen pollution from aquaculture be reduced? *NAGA, ICLARM*, 21: 17–25.

Priyanka and Garg, S.K., 2002. Replacement of fishmeal with canola (rapeseed) to enhance growth performance in the catfish *Heteropneustes fossilis* (Bloch.). *Environment and Ecology*, 20(3): 732–736.

Robinson Edwin, H. and Menghe, H.Li., 2007. *Catfish Protein Nutrition (Revised)*. Bulletin 1153. Office of Agricultural Communications, Mississippi State University, USA.

Singh, K., Garg, S.K., Kalla, Alok and Bhatnagar, Anita, 2003. Oil cakes as protein source in supplementary diets for the growth of *Cirrhinus mrigala* (Ham.) fingerlings. Laboratory and field studies. *Bioresource Technology*, 86: 283–291.

Steffens, W., 1989. *Principles of Fish Nutrition*. Ellis Horwood, Chichester.

Vielman, J., Makinen, T., Ekholm, P. and Koskela, J., 2000. Influence of dietary soybean and phytase levels on performance and body composition of large rainbow trout (*Oncorhynchus mykiss*) and algal availability of phosphorous load. *Aquaculture*, 183: 349–362.

Viola, S. and Lahav, E., 1993. The protein sparing effect of supplemental lysine in practical carp feeds. In: *Fish Nutrition in Practice*, (Eds.) S.J. Kaushik and P. Luquet. Paris–France–Institut.–National–De–La–Recherche–Agronomique.

Chapter 7

Ornamental Fish Packing and Health Management

☆ *Sachin Satam and Balaji Chaudhari*

Introduction

Fish and shellfish are transported in sealed plastic bags containing small quantities of water and pure oxygen. About 1.5-2 gallons of conditioned water is filled in a large tailor made high-density polyethylene bags. A prescribed number of fishes are placed inside the bag, which varies as per the size. The bag is filled with pure oxygen. The bag is sealed and placed into an insulated container (Styrofoam) and finally into a cardboard box (shipping box), labelled and transported over long distance with in 24-36 hours.

This type of shipment has become a standard method of transporting in recent time for the live fishes, based on many years of trial and error methods applied by many traders involved in transporting of live fishes. The earlier method of transporting very small fish or fry in small quantities in large tanks had always brought mortalities, even in very short distances.

The fishes, which are meant to be transported, have to undergo a set of tests, treatments before it is given the health certificate and to undergo the acclimatization procedure for the transport. The health management of these fishes are of utmost importance for the trade as this will reduce the mortality of the fishes (death on arrival), increases customer confidence, avoid stringent quarantine regulations and eliminate the chances of trans-boundary diseases. This chapter will focus on techniques and principals for identifying the healthy fishes, maintaining them in optimum conditions, screen them for any parasites and to starve them before transport without causing much stress.

Harvest

The health management of the fishes to be packed starts from its harvest till it is packed and sent to its destination. Based on the order received from the customers and watching the market trends one has to purchase fishes from the breeder or farmers or some times form the traders. Select the farms, which practices very hygienic conditions for the growing of fishes. The harvesting should be done with soft cotton netted gears during the early morning or late evenings in order to reduce the temperature shock. Careful handling will minimize losses due to stress and physical damage during harvesting. Overall, collecting and transporting fish should be made as quickly and carefully as possible to reduce potential risks in acquiring any injury, which later may aggravate to diseases.

Precautions to be Taken During Harvest

- ✰ The over crowding of fish during harvest, can rapidly deplete the oxygen supply in a localized area. This sub lethal oxygen stress often leads to later disease outbreaks. In order to avoid such conditions try to transfer fishes into holding buckets with good aeration and fresh and clean water. Note that in warm water oxygen is less soluble and fish respiration is higher.
- ✰ The protective slime of the fish is lost during handling at the time of harvest, thereby reducing their natural defence against pathogens (bacteria, fungus and viruses). Some times the loss of scales or cuts are direct cause for infection or mortality. The injuries can be minimised using soft cotton nets during harvesting and reduce the drag time so that the fishes are not in contact with the gear for long.
- ✰ Sudden change in water temperature is known to be a cause of fish stress. Sudden pH shock can also be harmful, especially to young fish. Harvesting during early morning or during evening will reduce the temperature stress. If the source water is not from the farm care should be taken to ascertain its pH before stocking fish in it, to avoid pH stress,.
- ✰ Always try to transport fish during the cooler parts of the day, and avoid long exposure to strong sunlight (Mid day). Do not expose fish to sun if they are in small containers, as the water will warm quickly.
- ✰ Care must be taken to provide the fishes with aeration or pure oxygen while carrying it to the packing centres in small polyethylene covers so as to help prevent oxygen stress and allow more fish to be safely transported. Also ensure that the water in the cover is kept cool and clean. Some medications may be added to the water to reduce risks of infection. Salt at 0.5 per cent is sometimes added to increase the tolerance of fish to handling.
- ✰ Always bear in mind that Small fish are usually more sensitive to handling and environmental stress than larger ones. For this reason, special precautions should be taken when transporting or stocking fry.

Packaging Centre

Once the fishes arrive at the packaging centre some important observations and measures have to be taken before they are stocked at the quarantine section in the designated tanks. The observations have to be made on the level and type of stress the fishes have undergone before it has reached the packaging centre, the stress may be,

Chemical Stress

- ✰ Low dissolved oxygen will make the fish lethargic and they tend to be inactive. They come to the top of the water column and try to gasp continuously; immediate supply of oxygen will help.

- ✩ The fishes which have been grown in the polluted water will have a stunted growth and will have different colour pattern, slime layer will be uneven, fungal patches will be found on the fins, some symptoms of fin rot may also be seen. Reject the stock.
- ✩ The fishes grown with out proper diets will have lean body, improper colour pattern. Weak fins and unhealthy appearance. Reject the stock.
- ✩ The fishes grown under improper water management or were the ammonia level might have been high, will be very weak and may die if little stress is applied.

Physical Stress

- ✩ Temperature is an important factor, which has to be carefully monitored during transport, or holding as this has a profound effect on the fish, it will kill the fish if it is sudden, either to higher or lower range. If a bag of fishes with good stocking density and oxygen arrive with more than 50 per cent dead then a major cause could be temperature shock.
- ✩ Always try to avoid exposing fishes to bright light because this may cause the fish to jump out of water, some times a hard hit on the walls of the tank may injure it. Over excitation may deplete the oxygen. Make sure the observation room does not have bright light.
- ✩ Some times air from the aerators is used to pack the fish; this will have high level of nitrogen in it, which may cause bubble disease. In some cases the air in the bag will accumulate carbondioxide, which may bring down the oxygen level in the water and make the fish to gasp at the surface. Always try to ascertain the level of stress the fish has undergone before taking any decisions.
- ✩ Always try to avoid exposing the fish to high level of vibrations or sounds this will make them inactive.

Biological Stress

- ✩ Careful packing means putting the right number of fishes in one packing considering its weight, character (especially fighter) and length groups. Always select packets, which are having less number of fishes.
- ✩ The behaviour of fish like aggression, territoriality, and lateral swimming space requirements plays a major role while considering it for packing. Fish packed in individual pack should be considered.
- ✩ When fishes are exposed to pathogens (Bacterial, Fungal or Virus); which are harmful and may kill the fish if not properly treated. Proper observation will reveal the symptoms like fin rot, fungal patches on the fins, swimming with jerks, losing the balance while swimming. Reject the stock.
- ✩ They may have external parasites (different types of Leech), and worms internally these parasites can be eliminated by giving treatment.

Procedural Stress

- ✩ The fish usually gets injured when they are not handled with right type of gear for a short duration or for a long duration. This inflicts injury on the body. Fish with minor injury may be treated but loss of scales or cut wound requires long time for getting cured.

- ☆ The means of transport plays an important role in the general health of fish in confined areas. The travel should not be jerky as it injures the fish, hot days should be avoided, always try to keep the temperature of the box constant and make sure that the travel is short. Any box arriving with more than 50 per cent dead should be rejected or put in treatment area and give proper treatment for a good recovery.
- ☆ A bag of fish may have 50–100 specimens, all the fishes may not in good healthy condition, this may happen if proper screening has not been carried out, this may cause trans – boundary disease. So quarantine is a must for all the batches at all times.

Scales and Skin

- ☆ Handling Stress most commonly damages Scales and skin. Any break in the skin, or removed scale, creates an opening for invasion by pathogenic organisms.
- ☆ Trauma caused by fighting (Reproductive Stress or Behavioural Stress) could result in breaks in the skin or scale loss.
- ☆ Parasite infestations can result in damage to gills, skin, fins, and loss of scales, which could create, breaks in the skin for bacteria to enter. Many times, fish which are heavily parasitized actually die from bacterial infections; but the parasite problem, associated physical damage, and stress response create a situation which allow the bacteria in the water to invade the fish, causing a lethal disease.

Prevention of Disease

- ☆ Fish farm management should be designed to minimize stress on fish in order to decrease the occurrence of disease outbreaks. When disease outbreaks occur, investigate the cause of mortality and identify all the biological, chemical or physical factors, which may be compromising the natural survival mechanisms of the fish. Correction of stressors (*i.e.* poor water quality, excessive crowding, etc.) should precede or accompany disease treatments.
- ☆ Stress breaks the fish's natural defences so that it cannot effectively protect itself from invading pathogens. A disease treatment is an artificial way of slowing down the invading pathogen so that the fish has time to defend itself with an immune response. Any stress, which adversely affects the ability of the fish to protect itself, will result in an ongoing disease problem; as soon as the treatment wears off, the pathogen can build up its numbers and attack again. Rarely would a treatment result in total annihilation of an invading organism. Disease control is dependent upon the ability of the fish to overcome infection as well as the efficacy of the chemical or antibiotic used.

The above are some of the stress a fish undergo when it is brought from the farmer, breeder or trader. This does not mean that such fishes should not used for trade, but this clearly shows that minimum handling means minimum mortalities and minimum losses. An ideal system should be

Fish ⟶ Quarantine ⟶ Breeding section ⟶ Grow out section ⟶

Packaging Centre ⟶ Export

all the above should be in a single place thus this will reduces the stress level on the fish, this may also reduce the amount of manpower, energy and time. This type of system will also provide the bio security for the ornamental fish.

Precautions to be Taken to Avoid Stress

- ✰ The keys to minimize disease outbreaks on your fish farm are maintenance of good water quality.
- ✰ Proper nutrition and sanitation.
- ✰ Prevention of disease outbreaks is more rewarding and cost-effective than treatment of dying fish.
- ✰ Disease treatments should never be applied in a haphazard fashion.
- ✰ When needed, chemical or antibiotic treatment should be targeted at a specific problem.
- ✰ Any management deficiencies in water quality management, nutritional management, or sanitation should be corrected.
- ✰ Fish, which do not respond to a correctly administered treatment, should be re-evaluated by a fish health professional.
- ✰ Experience and observation are the best teachers when learning how to avoid giving stress to the fish during handling and transport.
- ✰ Since the outcome of poor handling is generally sick or dying fish it does not make sense to take chances.
- ✰ Establish and follow a set of procedures, which minimize stress, and risk of injury to fish when they are handled.
- ✰ Use appropriate preventative treatments to ward off disease outbreaks before they occur.
- ✰ Look for patterns in the type and extent of diseases, which occur when compared with the type of handling the fish have received and under what conditions.
- ✰ Placing an emphasis on following these useful guidelines should help prevent disease problems and unnecessary losses.

Grading Orramental Fish

After the selection of the right type of stock, which has arrived, we have to ensure that quality product reaches the buyer. Quality is determined by a number of desirable characteristics, which include size, colour, fin development, sex. Certain anatomical observations have to be made such as scale loss, torn fin rays, eye damage, and parasite load, which will guarantee healthy animals. These procedures are time-consuming and this selection process is known as grading.

During grading, the packaging expert must keep in mind the quality requirements set by each buyer. The first criteria for grading should be the size (*e.g.*, small, medium, or large). This can be done visually by hand or with a mechanical grader. Grading fish mechanically allows the farmer to select fish of one size range in a short duration and accurately; as opposed to one at a time by hand. The colour is the next important criteria. Grading for this is achieved by hand only, if other characteristics like fin shape, body shape and eye size is considered each of it has to visually evaluated and separated. Careful handling during grading is important for minimizing stress and maximizing the quality of harvested fish. The amount of time spent grading fish is dependent on available manpower (number and skill) and equipment as well as the specific characteristics that are being.

Precautions to be Taken During Grading

- ✰ Grader must be gently agitated in the water more frequently to remove fish because openings of insert are only along the bottom and if too many big fishes are there then the small fishes will get struck in between and may be stressed.

- ☆ If adjustable grading machine are used then there is all possibilities that slats along side of the grader get caught up in the protective rubber siding.
- ☆ Usually the whole procedure is usually very rough on some of the sensitive fish.
- ☆ More time consuming and may still be necessary if grading for other characteristics besides size. It will be better that the manual grading is done using soft gears rather than putting the fish out of the water, as the stress factor will be very high.
- ☆ When the grading is done on the grading table it will be more stressful depending on the amount of time fish are held out of the water.
- ☆ Some experienced personnel should do the grading on table otherwise less experienced personnel may actually do more damage.

Prophylactic Treatments

Fish harvested from ponds and moved to indoor holding tanks often carry parasites and disease organisms with them. As a result of the stress of handling and the crowding of fish into tanks, the fish's resistance is lowered and a disease outbreak may occur. As a preventative measure, fish are often given a prophylactic treatment shortly after they have been brought in. Common treatments include a 2 hour bath of potassium permanganate at 3ppm (mg/l), or a 1 hour formalin bath at 50-100 ppm (mg/l). These treatments will eliminate most external parasites; the use of a prophylactic or preventative treatment may decrease the outbreaks of disease while fish are recovering from harvesting and handling and increase their survival during shipment. When using them for the first time it is important to be certain the dosage is not too stressful to the fish. If a bacterial disease is anticipated, bath treatments with antibiotics for 4 to 8 hours can be used. Most antibiotics are banned for use on food fish but several useful ones remain available to ornamental fish farmers.

The Treatment Strategy

- ☆ Salt (NaCl) used as a 0.5 per cent–1.0 per cent solution as an osmoregulatory aid or a 3 per cent solution for 10-30 minutes as a parasiticide.
- ☆ Formalin (37 per cent) is used as antiparasitic–(ecto) which is used 1 ml/10 gal for 11.5 hours.
- ☆ Nitrofurazone is an antibacterial, which is used 10 mg/litre in this fish, are treated for 11.5 hours.
- ☆ Potassium permanganate is an antiparasitic–(ecto) as well as antimicrobial, which is used 2 mg/litre for 4 hours.
- ☆ Quick Cure (99.2 per cent formalin + 0.75 per cent malachite green mix) is an antiparasitic–(ecto); especially for *Ichthyopthirius multifilis* ("Ich") and an antifungal, which is, used in the concentration of 0.5 ml/10 gal the fishes has to be dipped for 11.5 hours.
- ☆ Oxytetracycline is an antibacterial, which is added 1 g/10 gal in this fish is dipped for 11.5 hours.

The use of the above said drugs or chemical significantly improved the overall appearance and behaviour of fish over time, but these should be used under strict supervision and with out giving other physical, chemical or biological stress.

Precautions to be Taken

- ☆ Occasionally, as a result of severe stress during harvest or transport, fish will become sick and not respond to treatment. If this appears to be the case, the best solution is often to return the fish to the pond. Sometimes, although losses will occur, many fish will recover in the pond.

- ✰ Some chemicals (*e.g.*, potassium permanganate, formalin) may damage fish gills or skin and should be used carefully and only when absolutely necessary.
- ✰ Chemicals may also adversely affect water quality. For example, formalin decreases dissolved oxygen concentrations in water.
- ✰ Ideally, the Packing expert should be familiar with common health problems, actively work to identify and treat them as they become apparent, and seek assistance from a fish health professional when needed.
- ✰ Misuse of oxytetracycline or any other antibiotic, may lead to bacterial resistance making the drug ineffective.
- ✰ Before using any antibiotic, confirmation of a bacterial infection should be established and proper regimes should be followed.
- ✰ Note that oxytetracycline binds to calcium ions in the water, resulting in the need for a higher dosage of oxytetracycline in hard water [Total Hardness (TH) > 120 ppm].

Starvation and Conditioning Before Packing

The above said procedure may take few days and once the fishes are finalized for the packing it is starved for 24 hours to reduce the ammonia load during the transit. The second most important thing an expert packager should take care is the conditioning of the fish to the buyers' specification so that minimum loss occurs to the buyer while he is acclimatizing the fish to his water conditions this also guarantees the credibility of the seller.

Starvation is a form of stress so proper care has to be taken to note the length of the starvation so that the fishes can be relieved from this as soon as it reaches its destination this will further improve the response of the fish to the new environment.

Conditioning for a wide range of water parameters should be avoided as this may cause undue stress to the fish, which may lead to poor DOA. To some extent the conditioning should be done over a long period of time if it involves too many parameters to be changed.

Water Quality during Shipping

While in the plastic bags during the transportation process, fish health will be affected by changes in various water quality parameters. The parameters to be considered are temperature, dissolved oxygen, pH, carbon dioxide, ammonia, and the salt balance of the fish's blood. The rate of change of each parameter will be affected by the weight and size of fish to be transported and the duration of transport. Each water quality parameter will be discussed and methods of delaying the negative effects will be presented.

Temperature

Fish are cold-blooded; as a result, the temperature of their environment will affect the metabolic rate of fish. The metabolic rate of fish will double for each 18°F increase in temperature and be reduced by one-half for each 18°F decrease in temperature. A reduced metabolic rate will decrease the oxygen consumption, ammonia production, and carbon dioxide production. It is, therefore, essential to transport fish at low temperatures. For warm water species a temperature of 55°-60°F is recommended.

Coldwater fish such as trout inhabit colder water and naturally should be transported at even colder temperatures such as 45°-50°F.

To achieve the desired transport temperature, fish should be held in tanks that have access to cool water. By holding the fish in tanks for two days, the water temperature can be gradually reduced with additions of cool water from the cleanest available source. After loading the fish into bags final decreases and maintenance of temperatures during transport can be accomplished through additions of ice, or more commonly with the use of blue ice packs.

Ice or the blue ice packs are often used during transport, especially over longer transport periods that might allow increases in temperature. One-half pound of ice will reduce the temperature of one gallon of water by about 10°F. Insulated Styrofoam shipping boxes are also used to prevent outside temperatures from influencing the temperature of the transport water. In certain instances, ice coolers are used for transport.

Dissolved Oxygen

The most important single factor in transporting fish is the provision of adequate concentrations of dissolved oxygen (DO). The importance of supplying adequate levels of dissolved oxygen cannot be overemphasized. Failure to do so results in severe stress and possibly hypoxia or builds up of blood lactic acid, which may contribute, to fish kills two to three days after stocking.

The amount of oxygen that can be dissolved in water is based on water temperature. When the upper level is reached the water is referred to as being "saturated with oxygen." DO saturation is higher for cool water than for warm water. For example, at sea level DO saturation of 45°F water is 12.1 parts per million (ppm) while at 60°F saturation is 10.0 ppm. Because pure oxygen is used during bag transport, DO levels in the water will be saturated and the low oxygen levels will usually not be a problem, unless the bag is improperly sealed or develops holes caused from the spines of large fish. It is important to have a 75 per cent volume of oxygen in the bag to insure adequate diffusion of oxygen at the surface of the water.

pH

The quantity of hydrogen ions (H^+) in water will determine if it is acidic or basic. The scale for measuring the degree of acidity is called the pH scale, which ranges from 1 to 14. A value of 7 is considered neutral, neither acidic nor basic; values below 7 are considered acidic; above 7 basic. The acceptable range for fish growth is between pH 6.5-9.0. The pH of water will be influenced by the alkalinity (buffering capacity) and the amount of free carbon dioxide. The pH of the transport water will also affect the toxicity of ammonia. Even in well-buffered transport water the pH will sometimes decrease by one pH unit.

Carbon Dioxide

As fish respire they produce carbon dioxide as a by-product of respiration. Carbon dioxide will react with water to form a weak acid. This weak acid will in turn decrease the pH of the water. High levels of carbon dioxide (greater than 20 ppm) will interfere with the oxygen uptake in the fish's blood. High levels of carbon dioxide are sometimes found in well water. Excess carbon dioxide in well water can be reduced through mechanical aeration or by passing the water through a degassing column.

Ammonia

Ammonia build up occurs in transport water as a result of fish metabolism and bacterial action on fish wastes excreted into the water. Two forms of ammonia occur in transport water, ionized and un-ionized. The un-ionized form of ammonia (NH_3) is extremely toxic while the ionized form (NH_4+) is not. In tests for ammonia, both forms are grouped together as "total ammonia". The per cent of ammonia that is un-ionized will depend on both temperature and pH (Table 7.1).

Table 7.1: Per cent of Ammonia in the Un-ionized form at Different Temperatures (°F) and pH Values

pH	*Temperature*				
	50	*55*	*60*	*65*	*70*
6.0	0.02	0.02	0.03	0.03	0.04
6.5	0.06	0.07	0.09	0.11	0.17
7.0	0.19	0.24	0.29	0.34	0.43
7.5	0.59	0.74	0.93	1.07	1.33
8.0	1.83	2.30	2.87	3.31	4.10
8.5	5.56	6.92	8.54	9.78	11.90
9.0	15.70	19.00	22.80	25.50	29.90

Total ammonia concentrations may reach more than 14 ppm during transport. However, the per cent of the total ammonia, which is un-ionized at pH 6.5 and 55°F, is 0.07 per cent. Therefore, un-ionized ammonia at 14 ppm is 14 × 0.0007 = 0.0098 ppm. It is recommended that total ammonia concentrations greater than 5 ppm (0.015 ppm un-ionized at 60°F and pH of 7.0) be viewed with caution.

The easiest way to reduce toxic ammonia build up in transport water is to lower the temperature of the transport water and to stop feeding several days before transporting. Fish up to 8 inches should not be fed for 48 hours before loading and transporting and those larger than 8 inches should be off feed 72 hours before transporting.

Chemical Additives

Numerous chemical additives can be added to the transport water to alleviate several problems associated with transporting fish in sealed bags. Because overdoses of chemicals can result in death, care must be taken when measuring the dosage of each chemical used. It is essential to double-check every calculation and to use an accurate balance before adding chemicals.

The most common chemical added to transport water is salt (NaCl). Salt is used to relieve stress associated with maintaining a water balance in the fish. Fish have a blood salt concentration higher than the salts of the transport water. Concentrations of 5,000 ppm (0.5 per cent) are commonly used. A 5,000 ppm concentration can be made by adding 19 grams of salt per gallon (g/gal.) to water used during transport. The type of salt to use should be non-iodized containing no anti-caking compounds. Canning salt is a good example.

If the water alkalinity of the transport water is less than 100 ppm, some type of buffering compound should be added to the water. Properly buffered water will help remove free carbon dioxide which causes drops in pH. Sodium bicarbonate (Na_2CO_3) is one of the fastest reacting buffers and should be added at a rate of 1 g/gal. of water.

Finally, because fish are transported in crowded conditions, stress will be placed on them. Sometimes a chemical anaesthetic may be beneficial by producing a light sedation. The only anaesthetic for food fish is Finquel (tricaine methanesulfonate). Finquel may be used at a rate of 0.1-0.5 g/gal. of water.

Carrying Capacity

The maximum weight of fish that can be safely transported within a given period of time is the carrying capacity. Carrying capacity depends on the duration of haul, water temperature, fish size, and fish species. If water quality conditions such as temperature, oxygen, carbon dioxide, alkalinity, and ammonia are constant, then carrying capacity will depend on the fish species. Fewer pounds of smaller fish than larger fish can be transported per gallon of water. It is important for first time shippers, or experienced shippers transporting new species, to run test batches before undertaking any large shipment.

Transport Procedure

Days before the actual loading and transporting is to occur, the shipper needs to determine the carrier to be used, time of departure, time of arrival, and shipping costs. This information needs to be communicated to the receiver well in advance of the shipping date. With proper pre-planning, the risks of unnecessary delays in delivery and pickup are avoided. It is also the responsibility of the receiver to contact the shipper in the event of any mortality, which may be the responsibility of the shipper. All loading should be planned to allow boxes to be shipped as soon after loading as possible.

Procedures for bag shipping of fish are given below:

1. Carefully add the proper weight of fish to 1.5-2 gal. of clean degassed water. Water contained in the bag needs to be within two degrees of the holding water. Any chemicals should be added at this time.
2. Bag is deflated to remove air. Bag is then re-filled with pure oxygen. Approximately 75 per cent of the volume in the bag should be oxygen.
3. Mouth of bag is tightly twisted and secured with heavy-duty rubber bands Heat sealing can also be used.
4. Bag is placed inside second bag containing a frozen blue ice pack and sealed with rubber bands.
5. Sealed bags are then placed inside cardboard shipping box and sealed. The shipping box must be clearly labelled, "Live Fish" and have the name and address of the shipper and receiver. For extended trips, which may experience extremes in heat or cold the bags may need to be placed inside a Styrofoam cooler before being added to the shipping box.

Proper handling of bagged fish after receiving is as important as pre-handling to insure high survival. Guidelines for post-shipping are as follows:

1. Bags should be floated unopened in a shaded area of the receiving water for 30 minutes to allow temperatures to equalize. Observe for mortalities.
2. Open bags and quickly add 2-3 gal. of receiving water to the bag.
3. Slowly pour fish into the receiving water.

Chapter 8

Brachyuran Crab Resources of the Little Andaman Islands, India

☆ *Maloy Kumar Sahu, M. Murugan, R. Balasubramanian, S. Ajmal Khan and L. Kannan*

ABSTRACT

The brachyuran crabs constitute one of the important groups of Decapod crustaceans. From the perusal of literature on crabs, it is evident that only limited information is available on the crab faunal diversity of the Little Andaman island, which is known to be quite rich and varied in marine wealth, attracting the attention of many zoologists since the middle of the nineteenth century. Considering the above importance, a field survey was conducted for the first time in the inshore waters of the east coast of the island to assess the crab resources in the Little Andaman waters using various gears and crafts. During the present investigation, 27 species of brachyuran crabs belonging to 19 genera and 13 families were recorded. Among them, Portunidae topped the list with 3 genera and 6 species followed by Ocypodidae, Xanthidae, Leucosidae, Grapsidae, Carpilidae, Plumnidae, Majidae, Calappidae, Mictyridae, Dromiidae, Dorippidae and Plagusiidae. Further studies on the west coast of the island may possibly reveal many more species that have not been previously reported from the Bay waters.

Introduction

Among the marine fauna, crustaceans occupy a major place, which include brachuyaran crabs, hermit crabs, shrimps, lobsters and stomatopods. Brachyruan crabs in the last suborder of the Decapod in the Phylum Arthropoda are the unique "side-walkers", having a colourful exoskeleton and a pair of massive chela. Most of the brachyuran crabs are economically important. They fill the local protein demand and their medicinal values are much discussed. A few of them are not commercially important but they play a dominant role in the marine food-web process. The feaces of these crabs consist of carbon, nitrogen, phosphorus and trace metals which form a rich food for other consumers (Kuraeuter, 1976). The adult crabs and their larvae are consumed by many predators and omnivorous fishes. Thus, these crabs and their larvae have a vital role to play in the transfer of energy through the food-chain (Macintost, 1984). Further, these crabs are always the chosen test animals since their tolerance is very wide. Burrowing members of this group are of immense use in recycling nutrients thereby enhancing the richness of the soil by "ploughing".

Studies relating to the taxonomy and systematics of Indian brachyuran crabs are limited, though numerous studies are being carried out in other areas of fisheries science. Information on the brachyuran crabs of the Little Andaman island is known to a lesser extent, but is not complete. The present study, therefore, envisages a detailed investigation on the brachyuran crabs of the Little Andaman island.

Material and Methods

Brachyuran crabs were collected from eight stations (Station 1-Hut Bay, Station 2–Navel area, Station 3–Harabindra Bay, Station 4–R K Pur, Station 5–Dugong Creek, Station 6–Buttler Bay, Station 7–Netaji Nagar, Station 8–Nanjappa Nagar), along the east coast of the Little Andaman island which is situated between latitudes 10°30′ and 10°55′ N and longitudes, 92°33′ and 92°37′ E (Figure 8.1). Generally, hand picking was adopted for the collection of brachyuran crabs in the intertidal and sub-tidal zones during the tides. Sampling was done by the quadrate method. A quadrate of 50 × 50 cm was placed for sampling at an interval of 5 m in a transit parallel to the shore for about 200 m. The number of animals, which fell inside the quadrate, were counted each time and the distribution has been expressed as individuals/m^2 (Reys, 1964).

Results and Discussion

During the present investigation, 27 species of brachyuran crabs belonging to 20 genera and 13 families were recorded from the eight stations of the Little Andaman Island (Table 8.1). The family Portunidae topped the list with 6 species (22.22 per cent), followed by Ocypodidae with 5 species (18.52 per cent), Xanthidae with 3 species (11.11 per cent), Majidae, Grapsidae and Leucosiidae, each with 2 species (7.41 per cent) and Calappidae, Carpilidae, Pilummnidae, Dorippidae, Drommidae, Plagusiidae and Mictyridae, each with 1 species (3.70 per cent).

In a total of 20 genera, *Thalamita* and *Uca* topped the list with 3 species each followed by *Portunus, Ocypode* and *Docka* with 2 species each and the rest of the genera with 1 species each. In the present study a total of 239 individuals of species of brachyuran crabs were recorded (Table 8.2).

Results of the analysis of biodiversity indices are given below.

Species Richness

The maximum number of species was recorded at station 5 (21 species) and the minimum, at station 6 (16 species) and the order of decrease in species richness was as follows.

Station 5 (21 species) > Station 1 = Station 3 = Station 4 (each 20 species) > Station 2 = Station 8 (each 18 species) > Station 7 (17 species) > Station 6 (16 species).

Abundance

The maximum of species abundance was recorded at stations 1, 4 and 5 (each 34 nos.) and the minimum was at station 7 (24 nos.) and the order of decrease was:

Stations 1, 4 and 5 (each 34 nos.) > Station 6 (30 nos.) > Stations 2 and 3 (each 28 nos.) > Station 8 (27 nos.) > Station 7 (24 nos.).

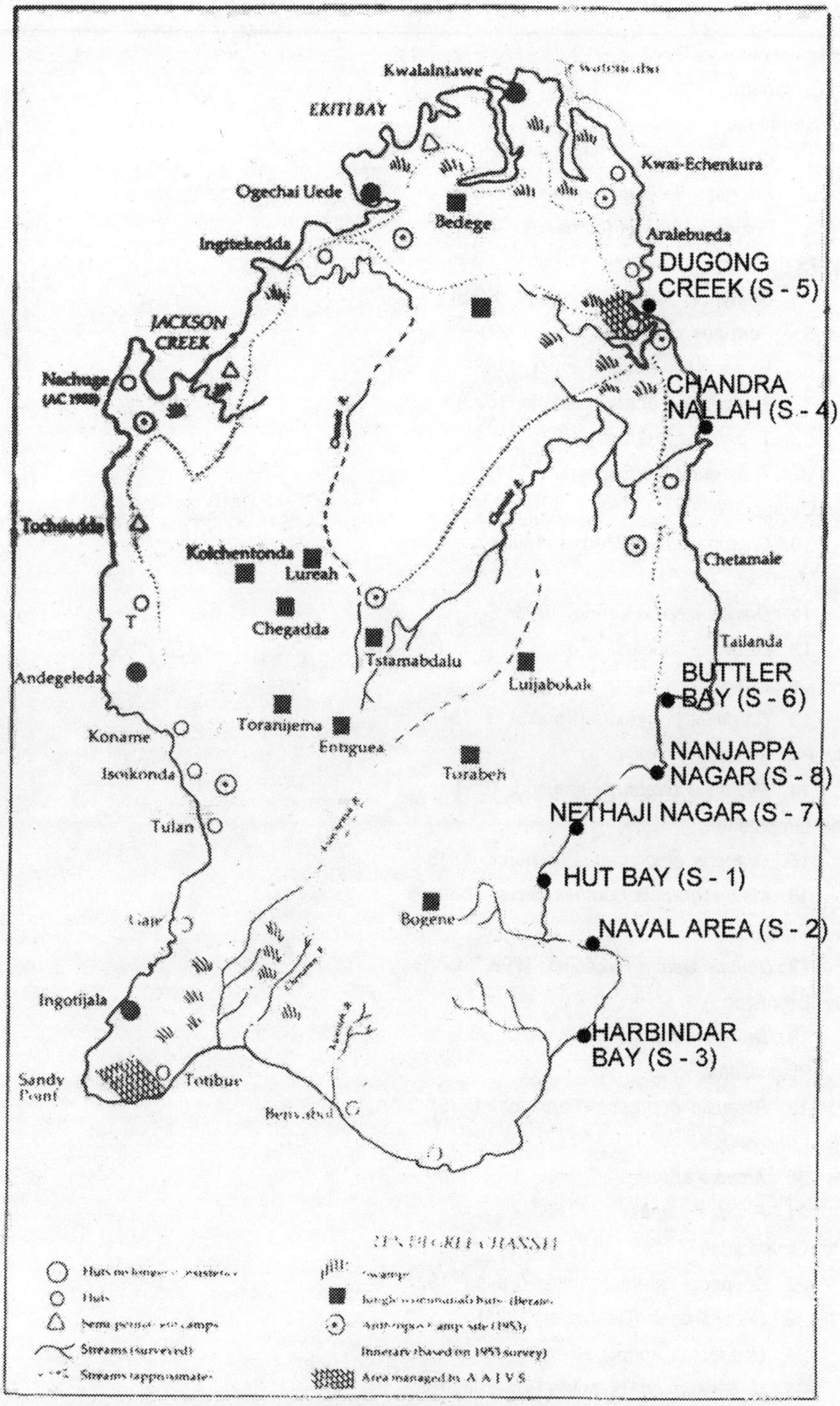

Figure 8.1: Map Showing the Study Areas in the Little Andaman Island, Along its East Coast

Table 8.1: Systematic List of Brachyuran Crabs Recorded from the Little Andaman Island

Class: Malacostraca

Order: Decapoda

Family: Xanthidae

1. *Galene bispinosa* (Herbst, 1783)
2. *Atergatis floridus* (Linnaeus, 1767)
3. *Zozimus aeneus* (Linnaeus, 1758)

Family: Portunidae

4. *Charybdis feriata* (Linnaeus, 1758)
5. *Portunus pelagicus* (Fox, 1924)
6. *P. sanguinolentus* (Herbst, 1783)
7. *Thalamita crenata* (Latreille, 1829)
8. *T. prymna* (Herbst, 1803)
9. *T. sima* (Milne Edwards, 1834)

Family: Calappidae

10. *Calappa gallus* (Herbst, 1803)

Family: Majidae

11. *Doclea alcocki* Laurie, 1906
12. *D. ovis* Fabricius, 1787

Family: Carpilidae

13. *Carpilius convexus* (Forskal, 1775)

Family: Pilumnidae

14. *Pilumnus hirsutus* Stimpson, 1858

Family: Grapsidae

15. *Grapsus albolineatus* Lamarck, 1818
16. *Metapograpsis frontalis* Miers, 1880

Family: Dorippidae

17. *Dorippe astuta* Fabricius, 1798

Family: Dromidae

18. *Dromia dehanni* (Rathbun)

Family: Plagusiidae

19. *Plagusia depressa* (Fabricius, 1775)

Family: Leucosiidae

20. *Arcania* sp.
21. *Randallia lanata*

Family: Ocypodidae

22. *Ocypode ceratophthalma* (Pallas, 1872)
23. *O cordimana* (Desmaret, 1825)
24. *Uca lactea annulipes* (H. Milne Edwards, 1837)
25. *U. dussumieri* (H. Milne Edwards, 1852)
26. *U. marionis* Ward, 1928

Family: Mictyridae

27. *Mictyris longicarpus* Latreille, 1806

Table 8.2: Quantitative Distribution of Brachyuran Crabs Recorded at the Different Stations of the Little Andaman Island (No/m²)

Sl.No.	*Species*	*St. 1*	*St. 2*	*St. 3*	*St. 4*	*St. 5*	*St. 6*	*St. 7*	*St.8*
	Family: Xanthidae								
1.	*Galene bispinosa*	2	1	0	3	1	4	2	1
2.	*Atergatis floridus*	1	2	1	0	3	1	0	3
3.	*Zozimus aeneus*	2	4	3	1	1	0	1	2
	Family: Portunidae								
4.	*Charybdis feriata*	2	0	2	2	2	1	0	1
5.	*Portunus pelagicus*	2	0	1	1	2	1	0	0
6.	*P. sanguinolentus*	1	0	1	2	0	2	1	2
7.	*Thalamita crenata*	4	1	1	0	1	0	1	0
8.	*T. prymna*	0	2	1	1	0	1	2	3
9.	*T. sima*	1	1	0	3	0	3	0	1
	Family: Calappidae								
10.	*Calappa gallus*	0	1	3	2	1	2	0	1
	Family: Majidae								
11.	*Doclea alcocki*	1	2	1	1	2	4	1	0
12.	*D. ovis*	2	3	0	1	3	0	0	1
	Family: Carpilidae								
13.	*Carpilius convexus*	0	1	0	3	1	2	0	1
	Family: Pilumnidae								
14.	*Pilumnus hirsutus*	2	0	1	1	1	0	2	1
	Family: Grapsidae								
15.	Grapsus albolineatus	1	1	2	0	2	0	1	2
16.	*Metapograpsis frontalis*	0	0	1	1	1	0	1	1
	Family: Dorippidae								
17.	*Dorippe astuta*	3	0	1	2	2	0	1	0
	Family: Dromidae								
18.	*Dromia dehanni*	0	1	2	0	1	0	1	1
	Family: Plagusiidae								
19.	*Plagusia depressa*	1	0	1	1	2	0	3	1
	Family: Leucosiidae								
20.	*Arcania*	1	1	2	2	0	1	2	0
21.	*Randallia lanata*	0	2	1	0	1	3	0	1
	Family: Ocypodidae								
22.	*Ocypode ceratophthalma*	1	1	0	1	2	1	0	0
23.	*O. cordimana*	0	2	1	0	1	0	1	1
24.	*Uca lactea annulipes*	2	0	1	1	1	1	2	0
25.	*U. dussumieri*	1	1	0	0	0	2	1	3
26.	*U. marionis*	2	1	1	3	0	1	0	0
	Family: Mictyridae								
27.	*Mictyris longicarpus*	2	0	0	2	3	0	1	0

Pielou's Evenness (J)

The maximum value was recorded at station 7 (0.970) and the minimum, at station 6 (0.942) and the order of decrease was:

Station 7 (0.970) > Station 5 (0.966) >Station 3 = Station 4 (each 0.965) > Station 1 (0.963) > Station 2 = Station 8 (each 0.957) > Station 6 (0.942).

Shannon-Wiener's Diversity (H)

The maximum value was recorded at station 5 (4.179) and the minimum, at station 6 (3.588) and the order of decrease was:

Station 5 (4.179) > Station 3 (4.027) >Station 1 (4.018) > Station 2 (3.995) > Station 4 (3.948) > Station 8 (3.913) > Station 7 (3.882) > Station 6 (3.588).

Simpson Index (D)

The maximum value was recorded at station 5 (0.939) and the minimum, at station 6 (0.905) and the order of decrease in abundance was:

Station 5 (0.939) > Station 1 (0.932) > Station 3 (0.931) > Station 2 = Station 4 (each 0.928) > Station 7 (0.926) > Station 8 (0.924) > Station 6 (0.905).

In the present study, 27 species of brachyuran crabs were recorded from the east coast of the Little Andaman island. Kariathil *et al.* (2002a) reported 16 species of brachyuran crabs belonging to 10 genera and 3 families from Campbell Bay, Great Nicobar Island where the family Ocypodidae and Grapsidae topped the list with 7 species each followed by Plagusiidae with 2 species. Moreover, the average population density of brachyuran crabs (15/m^2) reported by Dev Roy and Das (2000) from the Andaman waters were slightly higher than that of the present study. As compared to the previous works done in other Andaman and Nicobar groups of islands, this time, lesser brachyuran crabs species density and diversity have been noticed. The reason could be that the tsunami might have affected the coral reef area of this island, which is the major habitat for the brachyuran crabs and could have depleted the adult brachyuran crab populations, leading to the decrease of juveniles. Furthermore, after tsunami, the fishermen are not willing to go for fishing which has also led to the decrease of crab fishery landing in the island. However, further studies along the west coast of the island in future, may possibly add more species to the list of the Little Andaman that have not been previously reported from the Bay waters.

Acknowledgement

Authors thank Prof. T. Balasubramanian, Director, Centre of Advanced Study in Marine Biology and the authorities of Annamalai University for providing with necessary facilities. They also thank the Ministry of Environment and Forests, Government of India for financial support to carry out the work.

References

Deb, M. 1985a. A new genus and species of Portunid crab (Crustacea) from North Andaman. *Bull. Zool. Surv. India*, 7(2–3): 173–177.

Dev Roy, M.K. and Das, A.K., 2000. Habitat ecology. In: *Taxonomy, Ecobiology and Distribution Pattern of Brachyuran Crabs of Mangrove Ecosystem in Andaman Islands.*

Kariathil, T.J., Raffi, S.M., Khan, S.A. and Kannan, L., 2002a. Diversity, distribution and relative abundance of mangrove crabs in nullahs of Campbell Bay, Great Nicobar Island. In: *Proceedings of the National Seminar on Marine and Coastal Ecosystems: Coral and Mangrove: Problems and Management Strategies*, SDMRI Res. Publ., 2: 42–47.

Kariathil, T.J., Raffi, S.M., Khan, S.A. and Kannan, L., 2002b. Biodiversity, species composition, distribution and relative abundance of crabs in coral reef ecosystems of Campbell Bay, Great Nicobar Island. In: *Proceedings of the National Seminar on Marine and Coastal Ecosystems: Coral and Mangrove, Problems and Management Strategies*, SDMRI Res. Publ., 2: 125–131.

Kathirvel, M. and James, D.B., 1990. The phyllosoma larvae from Andaman and Nicobar waters. In: *Proceedings of the First Workshop on Scientific Results of FORV Sagar Sampada*, Cent. Mar. Fish. Res. Inst., Cochin, 147–150 pp.

Klinbunga, S., Siludjai, D., Wudthijinda, W., Tassanakajon, A., Jarayabhand, P. and Menasveta, P., 2001. Genetic heterogeneity of the Giant Tiger Shrimp (*Penaeus monodon*). *Mar. Biotechnol.*, 3(5): 428–438.

Kuraeuter, K.J., 1976. Biodeposition by salt marsh invertebrates. *Mar. Biol.*, 35: 215–223.

Macintost, D.J., 1984. Ecology and productivity of Malaysian mangrove crab population (Decapoda: Brachyuran). *Proc. As. Symp. Mangr. Env. Res. Manag.*, p. 354–377.

Reys, J.P., 1964. Les prelevements quantitatives du benthos de substratmeuble. *La Terre et la Vie.*, p. 94–105.

Chapter 9

Investigation on Tourism Effects of Macro Pollutants in the Beaches and Mangrove Environment, Southeast Coast of India

☆ *K. Balaji, S. Sudhakar, P. Raja, G. Thirumaran and P. Anantharaman*

ABSTRACT

The present investigation was carried out to evaluate the macro pollutants (solid wastes) of Vailankanni, Karaikal, Tranquebar, Poompukar, Pichavaram, Silver Beach-Cuddalore, Aurovil Beach-Pondicherry and Rocky Shore Beach-Pondicherry, Southeast coast of India during the period of February 2007 to July 2007. The level of solid waste contamination was higher in beaches, due to the coastal tourism. The major solid wastes were recorded in the present study includes cement/silica based wastes, fiber, glass wastes, metals, nylon wastes, paper, cloth materials, plastics, polythene wastes, rubber wastes, thermocole, sponges and leather wastes. The results were discussed in the relation to their possible impacts on the state of health of coastal flora and fauna.

Introduction

In India there are nine coastal states and two Union Territories. The maritime zones of the country are demarcated under the Maritime Zones Act 1976 as 12 nautical miles of territorial seas, 24 nautical miles of contiguous zone and 200 nautical miles of Exclusive Economic Zone (EEZ). India is one of the highly populated countries in the world. The developmental activities in the coastal zone, coupled with the population increase in the narrow stretch of land, pose enormous stress on the coastal environment, which affects the ecological balance such as, demographic pressure, rapid industrialization and urbanization in coastal cities and towns have added a variety of environment problems. The land based activities account for over 75 per cent of a marine pollution. The contaminants are transported to the coast through rivers, atmosphere and out falls. Thus the highly productive coastal areas lose their economic potential through pollution (Gowri and Ramachandran, 2001).

The seas in all parts of the world are littered with man-made debris, most of the plastics, which are particularly indestructible. The high concentration of plastic debris is found near busy shipping lanes and fishing areas. Discarded plastic items pose a special ecological problem because they are buoyant, durable and strong. Sea life is endangered by entanglement or ingestion. Plastic litter is conspicuous on many contemporary shorelines, most frequently near populated and industrial centers, but also on remote seldom visited or uninhabited Islands (Kartar *et al.*, 1973 and 1976).

Beaches are natural resources and their exploitations as a tourist attraction generates demand for services, jobs and income for local populations. Unplanned urban developments and population growth in urban centers close to the seashore lead to degradation of coastal environments through contamination by solid wastes. It pollute by solid waste is a global phenomenon. Studies related to solid waste contamination of beaches have been completed in many countries, and illustrate the diverse character of this sort of pollution, which includes plastics, nylon, polystyrene, organic debris, glass, metals and paper items (Garrity and Levings, 1993; Gregory, 1998; Whiting, 1998; Derraik, 2002; Araujo and Costa, 2003, 2004 a and b, 2005). Solid waste contaminants and accumulation of beaches are closely related to both human intervention and natural variables (Rees and Pond, 1995). The detritus doesn't just affect the beauty of our beaches; it is estimated globally over a million birds and 1,00,000 marine mammals and turtles die every year from entanglement, or ingestion of plastics (Laist, 1997).

Beachwatch is a nation-wide beach litter clean up and survey, organized by the Marine Conservation Society (MCS). It was launched in 1993 to raise awareness about marine and coastal litter, monitor levels and sources of litter on Britain's beaches and encourage action to reduce litter pollution at source. It is the flagship event of the MCS Adopt-a-Beach project, established in 1999, which encourages individuals; group communities to carry out regular seasonal litter surveys and tackle litter the local level (Marine Conservation Society–Beach watch, 2005). In order to bring the awareness on the marine litter and their detrimental effect in our marine environment, every year 20th September is observed as the International Coastal Clean up Day and more than 90 countries are participating in this one of the largest voluntary effort.

In this backdrop, several studies made on pollution in Indian coastal region in terms of heavy metal, pesticide and hydrocarbons. Unfortunately the literature on the solid waste pollution in coastal areas is less in India. Kaladharan *et al.*, (2004) investigates the occurrence of tarball and waste materials on the beaches along Kerala coast and Balaji *et al.*, (2006) surveyed on the macro pollutants in the intertidal and mangrove environment of Vellar estuary, Southeast coast of India. This investigation on solid waste is first of its kind in south east coast of India. In this present study, the survey of solid pollutants has been made from the recreational beaches of Vailankanni, Karaikal, Tranquebar, Poompukar, Pichavaram, Silver Beach-Cuddalore, Aurovil and Rocky Shore Beaches of Puducherry, Southeast coast of India during the period of February-July 2007. The level of solid waste contamination was higher in beaches, due to anthropogenic activities through coastal tourism.

Materials and Methods

Solid waste materials were collected as per the standard method (IOC, 1984) from one square meter quadrants, selected randomly within 1 km transect in intertidal zones of Vailankanni, Karaikal, Tranquebar, Poompukar, Pichavaram, Silver beach – Cuddalore, Aurovil and Rocky Shore Beaches of Puducherry. The materials collected from all the quadrants of transect were pooled together, sorted into plastic, rubber etc. and weighed.

Study Area

In the present study, the survey of solid pollutants have been made from the eight intertidal zones of Vailankanni (Lat. 10°41'; long 79° 51') is the world famous (Basilica church) holy place for Christians, which situated on the seashore, which is attracted by the more number of pilgrims from different parts of world. The Karaikal beach (Lat.10°54'; long 79°51') is one of the best natural beach in Puducherry (Union Territory). Now the Karaikal beach is attracting a heavy crowd of local people as well as of outside tourists. Tranquebar (Lat.11°02'; long 79°52') situated about 88 km South of Parangipettai and 300 km South of Chennai. The rocky shore habitat is formed by the damaged by the Masilamaninather Temple and old 17th centaury fort, which is submerged due to the encroachment of sea which attracts the near by local beach visitors and tourists from many other places.

Poompukar (Lat. 11°08'; Long. 79°51') has old heritage and formerly which is called as Cauary poompattinam which gather more tourists, Pichavaram (Lat. 11°27'; Long. 79°47') mangrove lies between the Vellar and Coleroon estuaries, is a heterogenous mixture of mangrove elements that spread over an area of 10 km^2. Silver beach – Cuddalore (Lat. 11°43'; Long. 79°49') is a sandy beach, which attracts more beach visitors from in and around areas, Aurovil Beach–Pondicherry (Lat. 11°02'; Long. 79°52') is located approximately 10 km by road north from the city of Pondicherry, which attracts more foreign beach visitors. Rocky Shore Beach – Pondicherry (Lat. 11°54'; Long. 79°50') is 180 km south from Chennai and 55 km north of Parangipettai. The seashore is characterized with rocks covering the steep sandy beach. These eight adjacent study areas are fall on southeast coast of India which is densely engaged with tourism activity.

Results and Discussion

In the present study the following contaminants were recorded aluminium screw cap, antibiotic capsule stripe, antibiotic oral suspension, asbestos sheet, audiocassette tape, balloon, biscuit pocket cover, broken glass, candy stick, canned food plastic cup, cement bags, chocolate rappers, cigarette/beddi butts, cigarette pocket, cool drinks bottle stickers and pockets, cycle tube, damaged fish nets, disposable plastic cups, dripper, electric wire, fiber boat pieces, flower ring, glass bottles, ice cream cup, incent stick pocket, iron pins, iron cap, leather cheppal, match box, match sticks, metal cans, milk pockets, mineral water bottles (Pet bottle), nylon materials, paint brush, paper, paper plates, piece of cloth, piece of cycle tyre, piece of plastic, piece of plastic paper, piece of plastic tap, piece of polythene, piece of pot, piece of PVC pipe, plastic ball, plastic box (different sizes), plastic buoyans, plastic carry bags, plastic cheppal, plastic comb, plastic cup, plastic spoon, polythene bags, polythene pouches, rubber ball, rubber cheppal, rubber tyre, shampoo pocket, sponge, straw, synthetic rubber cheppal, tea dust pocket, thermocole wastes and tube light/bulb during February-July 2007 in the eight stations and some of them are introduced into the sea through river runoff.

Table 9.1: Weight (g) of the Solid Wastes from throughout Study Periods

Station No.	*Stations*	*Total Solid Waste Weight (g/120 m²)*
1.	Vailankanni	59239.16
2.	Karaikal	43124.91
3.	Tranquebar	6097.2
4.	Poompukar	11977.36
5.	Pichavaram	16800.37
6.	Silver beach–Cuddalore	16725.24
7.	Aurovil Beach–Pondicherry	10353.98
8.	Rocky Shore Beach–Pondicherry	15364.44

The recorded pollutants were classified as follows cement/silica based wastes, fiber, glass wastes, metals, nylon wastes, paper, cloth waste materials, plastics, polythene wastes, rubber wastes, thermocole, sponges and leather wastes.

The present investigation recorded the total weight of macro pollutants (biodegradable and non-biodegradable) through out the study period along the beaches and mangrove environment of Coramandal coast are presented in Table 9.1. Our observations showed that Vailankanni (59239.16 g/120 m^2) beach were more contaminated than the other beaches. Because Vailankanni is a holy place, so it is attracted the more number of visitors whereas in Tranquebar beach (6097.2 g/120 m^2) is less polluted beach. Total numbers of solid wastes per month recorded from the study areas are shown in Figure 9.1. In the month of July shows the maximum level of contamination (208 numbers of solid wastes) in Vailankanni whereas in the month of March, April and May shows the minimum level of contamination (11 numbers of solid wastes) in Tranquebar and Poompukar. Hence the level of contamination is determined by the number of beach visitors.

The types of pollutants in numbers, weight and weight in percentage observed from beaches and mangrove environment are presented in Table 9.2a and 9.2b. Cloth wastes (percentage of weight 22.22 per cent) in Vailankanni, Nylon wastes (percentage of weight 30.99 per cent) in Karaikal, Plastic wastes (percentage of weight 26.78 per cent) in Tranquebar, Plant materials (percentage of weight 20.90 per cent) in Poompukar, Glass wastes (percentage of weight 26.30 per cent) in Pichavaram, Plastic wastes (percentage of weight 22.10 per cent) in Silver beach – Cuddalore, Plastic wastes (percentage of weight 38.81 per cent and 23.58 per cent) in Aurovil and Rocky Shore Beaches of Puducherry were observed as a high level pollutant.

Availability of dustbins and attention boards were observed from the study areas are presented in Table 9.3. Vailankanni is a densely engaged with tourism activity, but there is not avail any dustbins and attention boards. Among eight stations, in Tranquebar there is not avail any dustbins and attention boards. But level of contamination is also less. Because which is having less number of beach visitors.

Due to the litters in sea, they cause three kinds of impacts on the state of health of coastal flora and fauna such as entanglement, ingestion and toxicity. It is estimated that globally over a million birds and 1,00,000 marine mammals and turtles die every year. From our observations plastics are major pollutants, they are permanent additions to the sea but no environmental damage from them can be identified (Clark, 2001). 90 per cent of plastics are used as nest materials for gannet's on Grassholm Island (in the Bristol Channel) which indicating the high levels of plastic in the sea (Laist, 1997).

Table 9.2 a: The Types of Pollutants in Numbers, Weight and Weight in Percentage Observed from Beaches and Mangrove Environment

Sl.No.	Type of Solid Wastes	Stations											
		Vailankanni			Karaikal			Tranquebar			Poompukar		
		No.	Wt.	% of Wt.	No.	Wt.	% of Wt.	No.	Wt.	% of Wt.	No.	Wt.	% of Wt.
1.	Cement/Silica based wastes	0	0	0%	6	357	0.83	0	0	0	0	0	0
2.	Fiber wastes	11	1600	2.70%	4	1000	2.32	2	500.4	8.18	5	1250	10.44
3.	Glass wastes	30	5979	10.09%	25	6292	14.59	4	13.2	0.22	5	1523	12.72
4.	Metals	4	157.8	0.27%	11	177.8	0.41	1	2.3	0.04	0	0	0
5.	Nylon wastes	49	5593	9.44%	44	13367	30.99	15	1121	18.32	16	1624	13.56
6.	Paper wastes	125	1018.45	1.72%	98	514.4	1.19	17	145.8	2.38	22	218.4	1.82
7.	Cloth wastes	47	13160	22.22%	13	3159	7.33	4	1420	23.21	5	2500	20.87
8.	Plant materials	93	8039.3	13.57%	78	5034.9	11.67	16	5	0.08	12	2503.5	20.90
9.	Plastic wastes	251	10483.6	17.69%	118	8589.3	19.92	26	1638.3	26.78	28	1468.1	12.26
10.	Polythene wastes	186	1069.56	1.81%	128	731.11	1.70	36	125.1	2.04	52	197.16	1.65
11.	Rubber wastes	61	10059.7	16.98%	16	3130.8	7.26	2	480	7.85	2	243.6	2.03
12.	Thermocole wastes	50	311.25	0.53%	31	157.6	0.37	4	23.6	0.39	11	39.6	0.33
13.	Sponges	17	127.5	0.22%	0	0	0	1	7.5	0.12	0	0	0
14.	Leather wastes	8	1640	2.77%	3	615	1.43	3	615	10.05	2	410	3.42

Table 9.2b: The Types of Pollutants in Numbers, Weight and Weight in Percentage Observed from Beaches and Mangrove Environment

Sl.No.	Type of Solid Wastes	Stations											
		Pichavaram			Silver Beach–Cuddalore			Aurovil Beach–Pondicherry			Rocky Shore Beach–Pondicherry		
		No.	Wt.	% of Wt.	No.	Wt.	% of Wt.	No.	Wt.	% of Wt.	No.	Wt.	% of Wt.
1.	Cement/Silica based wastes	0	0	0	0	0	0	0	0	0	0	0	0
2.	Fiber wastes	7	1750	10.42	5	1250	7.47	10	13.5	0.13	6	1500	9.76
3.	Glass wastes	14	4418	26.30	9	2799	16.74	7	893.8	8.63	9	2602	16.94
4.	Metals	3	6.9	0.041	0	0	0	7	457.2	4.42	6	763	4.97
5.	Nylon wastes	7	2266	13.49	23	3546	21.20	5	125	1.21	11	1112	7.24
6.	Paper wastes	38	277.6	1.65	43	414.8	2.48	57	553	5.34	60	253.4	1.65
7.	Cloth wastes	8	536	3.19	0	0	0	0	0	0	0	0	0
8.	Plant materials	24	3508.5	20.88	25	2010.5	0.12	36	3015	29.12	51	2524.2	16.43
9.	Plastic wastes	61	2113.4	12.58	96	3696.83	22.10	76	4017.9	38.81	148	3622.5	23.58
10.	Polythene wastes	81	432.27	2.57	91	360.51	2.16	65	215.38	2.08	113	453.24	2.95
11.	Rubber wastes	4	1186.2	7.06	6	1960	11.72	2	186.2	1.80	7	1693.4	11.02

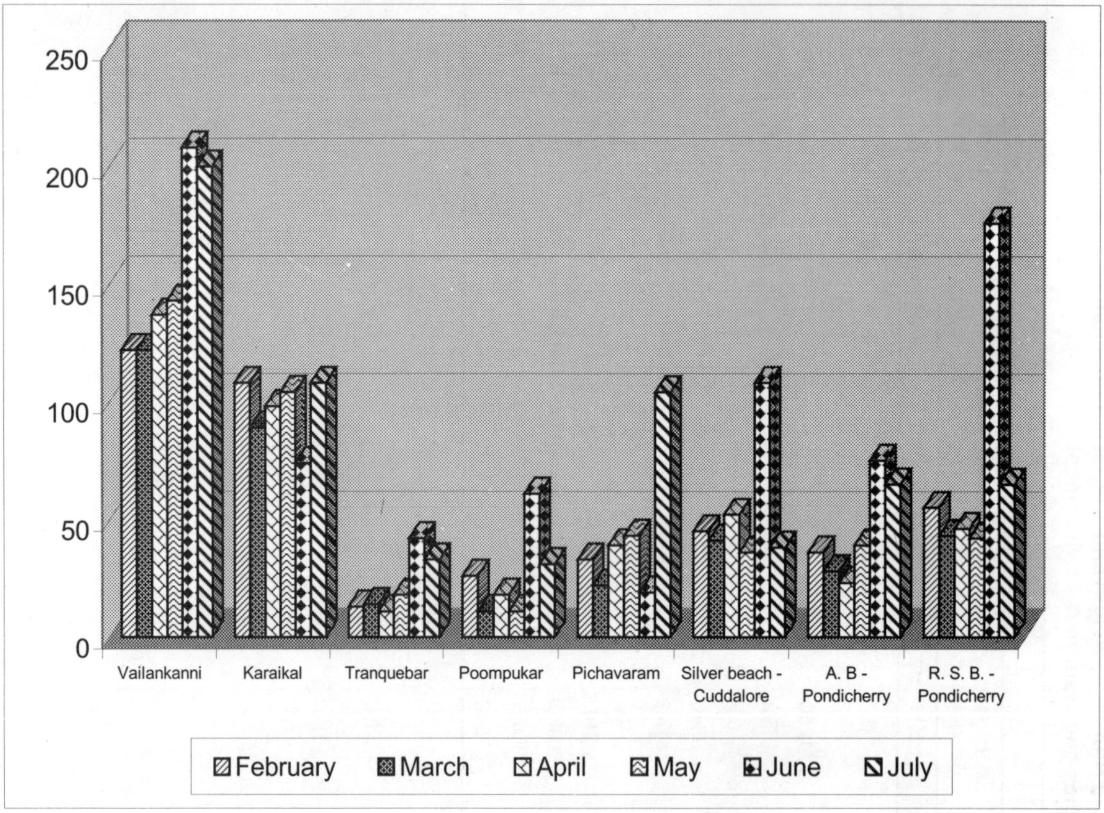

Figure 9.1: Number of Solid Wastes per Month

Table 9.3: Availability of Dustbins and Attention Boards in Study Areas

Station No.	*Stations*	*Availability of Dustbins*	*Attention boards*
1.	Vailankanni	Absent	Absent
2.	Karaikal	Absent	Absent
3.	Tranquebar	Absent	Absent
4.	Poompukar	Present	Present
5.	Pichavaram	Present	Present
6.	Silver beach–Cuddalore	Present	Present
7.	Aurovil Beach–Pondicherry	Absent	Absent
8.	Rocky Shore Beach–Pondicherry	Present	Present

Fowler (1995) studied that approximately 30,000 northern fur seals die annually due to entanglement primarily in net fragments. A survey of 25 grey seal pups off the Cornish coast found 17 to be oiled and four having plastic netting around their neck (Westcott *et al.*, 1994).

Ingestion of marine litter also occurs as items can be mistaken for food and damages the digestive tract leading to infection and starvation, and blocks the passage of food leading to death unless the blockage is removed. They may also result in the absorption of any toxic pollutants that were present on the debris (Third International Conference on Marine Debris, 1994). Turtles commonly mistakes a fine transparent plastic bag filled with water and debris looks a lot like a jellyfish, the favorite food. The Mediterranean Association to Save Sea Turtles (MEDASShiT) speaks of one sea turtle, which had ingested 50 such bags (Subhashini, 2004). A study of Fulmars (gull-like seabirds) from the Netherlands between 1982 and 2001 found that 96 per cent of the birds had plastic fragments in their stomachs, and in some cases this was the cause of death, reviewed by Van Franeker and Mejiboom, 2003 (Marine Conservation Society–Beach watch, 2005). In April 2002 a Minke whale was found stranded on a beach in Normandy. In the process of trying to determine its cause of death a stomach analysis found 800 g of plastic bags and packaging (http://perso.wanadoo.fr/gecc).

Burgess-Cassler *et al.*, (1991) stated large amounts of polyethylene and polypropylene, leaching pollutants can have implications for the biota. The accumulations of tiny plastic fibres in sand substrate may leak out toxins such as polychlorinated biphenyls (PCBs) and heavy metals (Thomson and Hoare, 1997) PCBs have been found to accumulate in top predators such as seals and dolphins.

Cigarette butts are thrown away by beach visitors after smoking which are prime solid wastes occurring in beaches and they remain in the environment for a long time as they contain a plastic called cellulose acetate. They leach toxic chemicals into the environment, and are mistaken for food by marine animals, which have been found in the stomachs of dolphins, sea birds, and turtles (Marine Conservation Society–Beach watch, 2005).

Accumulation of litter can affect many species, especially as some areas of the sea floor act as sinks for litter (Galgani *et al.*, 2000). Smothering can prevent light and important nutrients from reaching benthic organisms. Abrasion by heavy litter items can also cause substratum damage to the seabed (Laist, 1997). In addition, smaller processes can contribute to the effect of litter on the sea bed in (Iribarne *et al.*, 2000) the Southern West Atlantic, populated by the burrowing crab (*Chasmagnathus granulata*) there are significantly larger amounts of mostly plastic debris on the surface sediments and more buried debris in the sediments, than in areas not populated by the crab (Marine Conservation Society–Beach watch, 2005).

Debris, includes floating items and natural, may provide shelter and food for small fish and migrating animals. This is a method of transport for colonization and recent research has found that the occurrence of plastic marine debris on a global scale increasing the risk of invasion of new habitats by alien species.

The winds and currents can act directly in the transport of the wastes, especially those of low density such as plastics (Thornton and Jackson, 1998). The winds that blow over the beach environment are also responsible for solid waste deposition and excess accumulation and support our study, which was conducted in the intertidal beach; here the solid wastes are transported by the activity of tides.

Thus the non-degradable solid wastes such as plastics and polyethylene materials may be separated and reused or recycled and biodegradable materials may be utilized as manure for agriculture. The famous campaign-"Bag it and Bin it" (Gowri and Ramachandran, 2001) should be popularized among coastal population so that the garbage may be disposed of without causing damage to the environment. Municipal garbage may be used for power generation through gasification, about 800 metric tons of garbage is capable of generating about 15 MW power. Though the initial cost will be high, this project will be most useful in the long-term in reducing the pollution as well as in making use of solid wastes.

Conclusion

Tourism industry is an important economic source for foreign exchange, domestic product, income and employment of the nation. Even though, without causing pollution is a vital part to conserve the nature. Coastal and marine environment are increasingly threatened by increasing coastal tourism activity and also by many anthropological activities due to solid waste dumping along the rivers, creeks, estuaries, mangroves and coastal marshy lands. Hence in this context, the proper sustainable solid waste management was needed to reach the goal of "Preventing Marine Pollution" is unavoidable at this hour.

References

Araujo, M.C.B. and Costa, M., 2003. Lixo no ambiente marinho (Rubbish in the Marine Environment). *Ciencia Hoje*, 32: 64–67.

Araujo, M.C.B. and Costa, M., 2004a. Analise quail-quantitativa do lixo deixado na Baia de Tamandare, PE-Brasil por excursionistas (Quali-quantitative analysis of the waste left at Tamndare Bay, Pernambuco Brazil, by visitors). *Jounal de Gerenciamento Costeiro Integrado*, 3: 58–61.

Araujo, M.C.B. and Costa, M., 2004b. Quali-quantitative analysis of the solid wastes at Tamndare Bay, Pernambuco Brasil, *Tropical Oceanography*, 32: 159–170.

Araujo, M.C.B. and Costa, M., 2005. Municipal services on tourist beaches: Costs and benefits of solid wastes collection. *Journal of Coastal Research*, 21(2).

Balaji, K., Thirumaran, G., Kumaraguru Vasagam, K.P. and Anantharaman, P., 2006. A survey of the macro pollutants in the intertidal and mangrove environment of Vellar estuary, Southeast coast of India. *Ecol. Env. and Cons.*, 12(2): 199–202.

Beach watch, 2005. Campaign of the week: Beachwatch (Online) Available: http://www.adoptabeach.org.uk/LitterFacts/shipsolntions.htm.

Burgess-Cassler, A., Imam, S.H. and Gould, J.M., 1991. High molecular weight amylase activities from bacteria degrading starch plastic film. *Applied Environmental Microbiology*, 57: 612–614.

Clark, R.B., 2001. *Marine Pollution*, 5th Edn. Ford University Press Inc., New York, pp. 1–237.

Countryside Council for Wales, 1999. Conservation of Strandlines (booklet). CCW, Bangor, Gwynnedd.

Derraik, J.G.B., 2002. The pollution of the marine environment by plastic solid wastes: A review. *Marine Pollution Bulletin*, 44: 842–852.

Fowler, C.W., 1985. An evaluation of the role of entanglement in the population dynamics of northern fur seals on the Pribilof Islands. In: *Proceedings of a Workshop on the Fate and Impact of Marine Debris*, (Eds.) R.S. Shomura and H.O. Yoshida, 27–29 November, 1984, Hawaii, pp. 291–307.

Galgani, F., Leaute, J.P., Moguedet, P., Souplet, A., Verin, Y., Carpentier, A., Goraguer, H., Latrouite, D., Andral, B., Cadiou, Y., Mahe, J.C., Polard, J.C. and Nerisson, P., 2000. Litter on the sea floor along European coasts. *Marine Pollution Bulletin*, 40(6): 516–527.

Garrity, S.D. and Levings, S.C., 1993. Marine debris along the Caribbean coast of Panama. *Marine Pollution Bulletin*, 26: 317–324.

Gowri, V.S. and Ramachandran, S., 2001. Chapter–12 *Coastal Pollution Management*, Coastal Environment and Management. Institute for Ocean Management, pp. 176–194.

Gregory, M. R., 1998. Pelagic plastics and other synthetic marine solid wastes: A chronic problem. In: *Marine Ecosystem Management: Obligations and Opportunities*, (Eds.) C. Wallace, B. Weeber, S. Buchanan. ECO (Environment and Conservation Organizations of New Zealand), pp. 128–135.

IOC., 1984. *Manuals and Guides*, UNESCO, 13: 35.

Kartar, S., Milne, R.A. and Sainsbury, M., 1973. Polystrene waste in the Severn estuary. *Mar. Pollut. Bull.*, 4: 44.

Kartar, S., Milne, R.A. and Sainsbury, M., 1976. Polystrene spherules in the Severn estuary : A progress report. *Mar. Pollut. Bull.*, 7: 52.

Laist, D., 1997. Impacts of marine debris: Entanglement of marine life in marine debris including a comprehensive list species with entanglement and ingestion records. In: *Marine Debris: Sources, Impacts Solutions*, (Eds.) J. Coe and D.B. Rogers. Springer Series on Environmental Management. Marine Conservation Society, 2006. Marine Conservation Society Beachwatch, 2005, Executive Summary, pp. 1–7.

Rees, G. and Pond, K., 1995. Marine litter monitoring programmes :A Review of Methods with Special Reference to National Surveys. *Marine Pollution Bulletin*, 30(2): 103–108.

Third International Conference on Marine Debris, 1994. Draft final Working Group Reports. Miami, Florida, USA.

Thompson, R. and Hoare, C., 1997. Microscopic plastic: A Shore thing. *Marine Conservation*, 3(11).

Thornton, L. and Jackson, N.L., 1998. Spatial and temporal variations in debris accumulation and composition on a estuarine shoreline. Cliffwood Beach, New Jersey, USA. *Marine Pollution Bulletin*, 36(9): 705–711.

Westcott, S., Semmens, P. and Schofield, C., 1994. Marine Debris in Cornwall. Cornwall Wildlife Trust, UK.

Whiting, S.D., 1998. Types and sources of marine debris in Fog Bay, Northern Australia. *Marine Pollution Bulletin*, 36(11): 904–910.

Chapter 10

A Comparison of Live Feed and Supplementary Feed for the Growth of Catfish Fry, *Clarias batrachus* (Linn)

☆ *Meenakshi Jindal, N.K. Yadava and Manju Muwal*

ABSTRACT

A comparison was made on growth of *C. batrachus* fry by using three alternative feed sources during the early growth phase of fish, such as zooplankton as live feed, plant (soybean) and animal (fishmeal) based feed as supplementary feeds. Low dissolved oxygen (DO) values in fish groups fed on zooplanktons clearly indicate its utilization by the growing fry. Fish weight gain showed negative correlation with NH_4-N and O-PO_4 excretion indicating that high NH_4-N excretion repressed growth. Among all the food types, zooplankton was found to produce the best growth results followed by soybean based diet, while fishmeal based feed yielded poor results. Therefore, our finding strongly support the use of live feed during the early growth phase of *C. batrachus.*

Introduction

Air breathing catfish *Clarias batrachus,* locally known as magur, is of great demand and attracts the attention of farmers for its high market value. This species is very famous because it is an important part of the diet for children and lactating mothers. Furthermore, the species can be kept alive for long time by storing them in water container without giving any food as the species bear special accessory respiratory organ. This species is highly regarded for food due to its high protein (15.0 per cent), low fat (1.0 per cent) and high iron content (710mg/100g tissue).

The large-scale commercial spawning and culture of *C. batrachus* seeks a great demand on available live food at hatcheries. Various dry-feed formulae have been investigated as possible substitutes of live food for larval development (Dabrowski, 1984). In recent years, suitability of various dry-feed formulae has been investigated for the rearing of cyprinid and catfish larvae (Uys and Hecht, 1985). However, it has been shown that formulated compound diets do not provide optimal larval growth when used exclusively as larval food, especially during the early larval stages of cyprinids and catfish (Hoogendoorn, 1980; Dabrowski, 1984; Prinsloo and Schoonbee, 1986), therefore, live food organisms still remain as the major food source. Natural live food provides a substantial availability of protein and other essential nutrients (Nayak *et al.*, 2003).

In the present study, a comparison was made on the growth of *C. batrachus* using live feed (zooplanktons) and supplementary feed (fishmeal based and soybean based) sources during the early growth phase (fry) of this species. During the course of study, efforts were also made to study the excretory levels of ammonia and phosphorous excreted by the fry in the treated water.

Material and Methods

Mass Culture of Zooplankton

Zooplankton were cultured in glass aquaria of 100L capacity in the screen house of Department of Zoology and Aquaculture, CCS, HAU, Hisar in a replicate of three. These aquaria were filled with water and a mixture of manure (cattle manure: mustard oil cake: poultry in the ratio of 1:1:1) was added @ 200g/aquaria (Garg, 2003). After 10 days, zooplankton samples were collected from a fish pond, CCS HAU, Hisar and inoculated in these glass aquaria. After about 10 days, zooplankton from these aquaria were used for feeding the fry as a live feed.

Preparation of Supplementary Feed (Plant and Animal Based)

Groundnut oil cake, rice bran, wheat flour, fishmeal and soybean were procured from local market of Hisar. Soybean seeds contains some anti-nutritional factors such as lectins, phyto-haemoglutinins, trypsin inhibitors etc. which were eliminated by heat processing in an autoclave at 121°C, 15lb for half an hour (Garg *et al.*, 2002). All the ingredients were grinded, powdered and subjected to proximate analysis following AOAC (1995) (Table 10.1).

Table 10.1: Proximate Analysis (Per cent dry weight basis) of the Ingredients Prior to the Preparation of Supplementary Diets (Fishmeal based and soybean based)

Ingredients	*Proximate Composition (Per cent)*					
	Crude Protein	*Crude Fat*	*Crude Fiber*	*Total Ash*	*Nitrogen Free Extract*	*Gross Energy (KJg^{-1})*
Groundnut Oil Cake (GNOC)	35.266 ± 0.005	6.250 ± 0.003	6.000 ± 0.005	7.000 ± 0.000	45.496 ± 0.006	18.607 ± 0.003
Rice Bran (RB)	14.100 ± 0.005	10.066 ± 0.005	11.003 ± 0.005	20.556 ± 0.003	44.440 ± 0.005	14.906 ± 0.007
Wheat Flour (WF)	14.290± 0.029	2.666± 0.088	1.533± 0.033	1.366± 0.088	80.143± 0.407	18.177± 0.037
Fish Meal (FM)	42.926 ± 0.002	10.996 ± 0.003	3.493 ± 0.003	29.653 ± 0.003	12.930 ± 0.001	16.713 ± 0.000
Processed Soybean* (HPS)	43.733 ± 0.008	25.603 ± 0.000	4.496 ± 0.003	3.796 ± 0.000	22.370 ± 0.001	24.298 ± 0.001

* Raw soybeans were hydrothermically processed in an autoclave at 121°C at 15 lbs for 30 min. to remove anti-nutritional factors (ANFs) (Garg *et al.*, 2002).

\# All values are mean±S.E. of means of 3 observations.

Two types of supplementary feeds were prepared to feed the fry of *C. batrachus*. One is fishmeal (animal protein) based and second is soybean (plant protein) based. The ingredients content and proximate compositions of these two supplementary feeds was given in Table 10.2.

Table 10.2: Ingredient Contents and Proximate Composition of Two Supplementary Feeds Used for Feeding the Fry of *C. batrachus*

	Supplementary Feeds	
	FM Based	*Soybean Based*
Ingredients		
Groundnut oil cake (GNOC)[A1]	60	60
Rice bran[A2] (RB)	5	5
Wheat flour[B] (WF)	5	5
Fishmeal[C] (FM)	28	–
Processed soybean[D] (HPS)	–	28
Mineral premix and amino acids[E] (MPA)	1	1
Chromic Oxide[F] (Cr_2O_3)	1	1
Proximate composition (Per cent)		
Crude protein	40.25	40.25
Crude fat	7	9.5
Crude fiber	3.5	7.25
Total Ash	7.3	6.5
Nitrogen free-extract (NFE)	42.95	36.5
Gross energy KJ/g	19.65	19.53

A1 and A2: Used as basic feed ingredients; B: Used as a binder to make diets water stable; C: Used as supplementary feed of animal origin; D: Used as supplementary feed of plant origin; E: Used to supplement the diets with minerals and amino acids.

Each Kg contains Copper: 312mg; Cobalt: 45mg; Magnesium: 2.114g; Iron: 979mg; Zinc: 2.13g; Iodine: 156mg; DL-Methionine: 1.92g; L-lysine mono hydrochloride: 4.4g; Calcium: 30 per cent and Phosphorous: 8.25 per cent.

F: Used as external digestibility marker

Experimental Design

Fry of *C. batrachus* (mean body weight 0.71 to 0.76g) were procured from CIFE Centre Lahli, Rohtak during the month June. They were acclimatized in glass aquaria in the laboratory of Department of Zoology and Aquaculture, CCS, HAU, Hisar for a minimum period of 15 days prior to the commencement of experiment. The aquaria water was renewed daily with dechlorinated water. During acclimatization period, the fry were fed on supplementary feed (FM based and Soybean based) alternatively.

All treatments were conducted in transparent glass aquaria (60×30×30cm), containing 30L of chlorine free water in a replicate of three, kept in a temperature controlled laboratory. All groups of fry (each group containing 20 fry) were fed daily between 4:00 to 5:00 pm, @ 5 per cent BWd^{-1} for the whole experimental duration of 70 days. One group of fry were fed on live feed by transferring a measured quantity of water containing zooplanktons everyday for a period of 70 days. Growth of fry was monitored after regular day's interval in terms of weight and length and feeding rate adjusted accordingly. After 4hrs of feeding, the uneaten feed was siphoned out especially in case of supplementary feeds (FM and soybean based) otherwise they will deteriorate the water of aquaria; Individual weight of the fish fry was recorded at the beginning and at the end of the experimental period of 70 days.

Analytical Techniques

Live weight gain (g), length gain (cm), growth percent gain, specific growth rate (SGR) in terms of length and weight were calculated using standard methods given by Steffens (1989). Physico-chemical parameters like dissolved oxygen (mg/l), temperature (°C), pH, conductivity (μ mhos cm^{-1}), total alkalinity (mg/l), total hardness (mg/l), free CO_2 (mg/l) and excretory levels of ortho–phosphate ($O-PO_4$) and ammonical nitrogen (NH_4-N) mg $100g^{-1}$ BWd^{-1} of aquaria water of different treatments were recorded following APHA (1998).

Statistical Analysis

Data was analysed following ANOVA, at 5 per cent probability level. Group means were compared by student 't' test.

Results and Discussion

Mass Culture of Zooplankton

Various zooplankton observed during the mass culture is listed in Table 10.3. The data showed the population of protozoa, rotifers, crustacean, copepods and ostacoda during the mass culture. The data also showed that the population of Daphnia followed by Trinema were observed higher than other species of zooplankton.

Table 10.3: Population of Different Zooplanktons Observed During Mass Culture of Zooplanktons for a Period of 30 Days Under Semi Lab Conditions

Zooplankton	*Observation Days (Zooplankton Nos./ml)*					
	5th	*10th*	*15th*	*20th*	*25th*	*30th*
PROTOZOA						
Amoeba	1	6	9	11	7	9
Trinema	2	5	10	9	10	11
Paramoecium	1	3	3	5	6	7
Total protozoa	4	14	22	25	23	27
ROTIFERA						
Branchinous sp.	1	3	4	5	6	7
Rotaria	1	4	2	4	7	9
Total rotifera	2	7	6	9	13	16
CRUSTACEA–CLADOCERA						
Daphnia sp.	3	6	12	11	13	15
Moina sp.	2	4	5	7	8	10
Ceriodaphnia	1	2	3	5	6	6
Total	6	12	20	23	27	31

Contd...

Table 10.3–Contd...

Zooplankton	*Observation Days (Zooplankton Nos./ml)*					
	5th	*10th*	*15th*	*20th*	*25th*	*30th*
COPEPODA						
Cyclops sp.	1	3	4	6	7	9
Diaptomous sp.	0	2	3	5	7	10
Total copepoda	1	5	7	11	14	19
OSTRACODA						
Cypris	0	1	2	7	10	10
Total ostracoda	0	1	2	7	10	10
TOTAL ZOOPLANKTON	13	39	57	75	87	103

Water Quality Parameter

The data showing water quality parameters of different treatments are presented in Table 10.4. The groups of fry receiving zooplankton were the most affected with lowest oxygen (DO) concentration. This clearly showed that utilization of zooplankton by the growing fry. These results are in agreement with those of Nayak *et al.* (2003) and Kalla *et al.* (2004). Total alkalinity, hardness and electrical conductivity were found to be lower in case of fry fed on zooplankton rather than fry fed on supplementary feed (FM and soybean based).

Table 10.4: Periodical Analysis of Physico-chemical Parameter of Treated Diets (Live feeds and supplementary feeds) in Fish Laboratory for an Experimental Period of 70 Days

Sl.No.	*Parameters*	*Range*		
			Supplementary Feeds	
		Live Feeds	*FM Based*	*Soybean Based*
1	Temperature (°C)	26.0–30.0	26.0–30.5	26.5–30.0
2	Dissolved oxygen (mg/l)	5.4–6.4	5.5–6.7	5.2–6.3
3	pH	7.6–7.9	7.9–8.1	7.8–8.0
4	Alkalinity (mg/l)	245–285	252–290	249–283
5	Total hardness (mg/l)	205–230	217–241	210–232
6	Free CO_2 (mg/l)	15.6–17.1	5.4–17.2	15.2–16.8
7	Electrical conductivity (μ mhos/cm)	0.49–0.52	0.52–0.55	0.50–0.51
8	Ammonical Nitrogen (mg g^{-1} BW of fish)	0.170–0.244	0.179–0.250	0.172–0.246
9	Orthophosphate (mg g^{-1} BW of fish)	0.211–0.268	0.215–0.272	0.215–0.276

Post-Prandial Excretory Levels of Ammonical Nitrogen (NH_4-N) and Ortho-Phosphate (O-PO_4)

The ammonical nitrogen (NH_4-N) and ortho-phosphate (O-PO_4) concentration in water was found to be lower in zooplankton troughs in comparison to supplementary (dry feed) troughs as shown in Table 10.4. These results are in agreement with those of Nayak *et al.* (2003). The results further showed that among dry feeds, soybean based feed excrete less ammonia and phosphorus rather than FM based feed indicating that plant based proteins excrete less ammonia and phosphorus rather than animal based protein. These results are in agreement with those of Kalla *et al.* (2004); Jindal and Garg (2005); Jindal *et al.* (2007a, b) and Jindal (2008).

This indicated that unless proper cleaning and water exchange were done daily, the concentration of ammonia and phosphorus would have been much more, which would be detrimental to fry growth and may cause mass mortality of fry.

The peak in excretion of NH_4-N were observed at 8h of post–feeding (Figure 10.1) but the peak in the excretion of O-PO_4 was observed at 6h of post–feeding (Figure 10.2). These results are in line with those observed by Kalla *et al.* (2004), Jindal *et al.* (2007a,b) and Jindal (2008).

Growth and Survival

Fry mortality was very less and independent of experimental treatment. The growth response of the experimental fry on live feed and supplementary feeds (FM based and soybean based) are shown in Table 10.5.

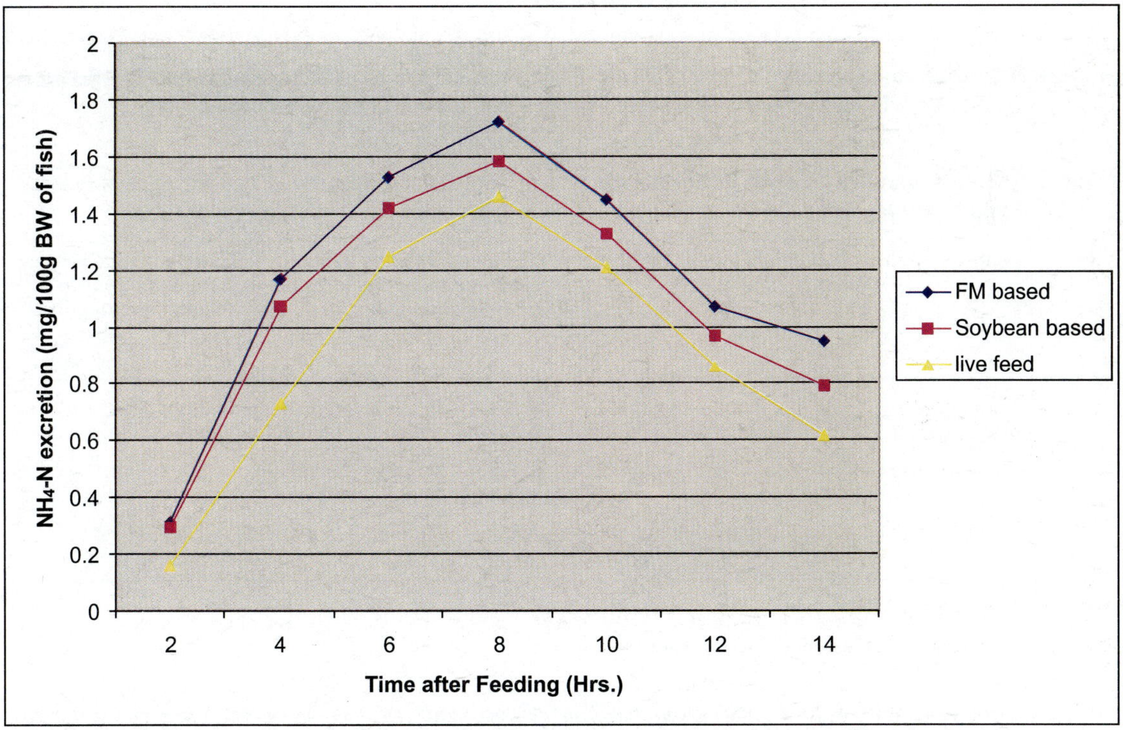

Figure 10.1: Diurnal Excretory Pattern of Ammonical Nitrogen (NH_4-N) in *Clarias batrachus* Fry Fed on Live Feed and Supplementary Feeds (CFM based and soybean based)

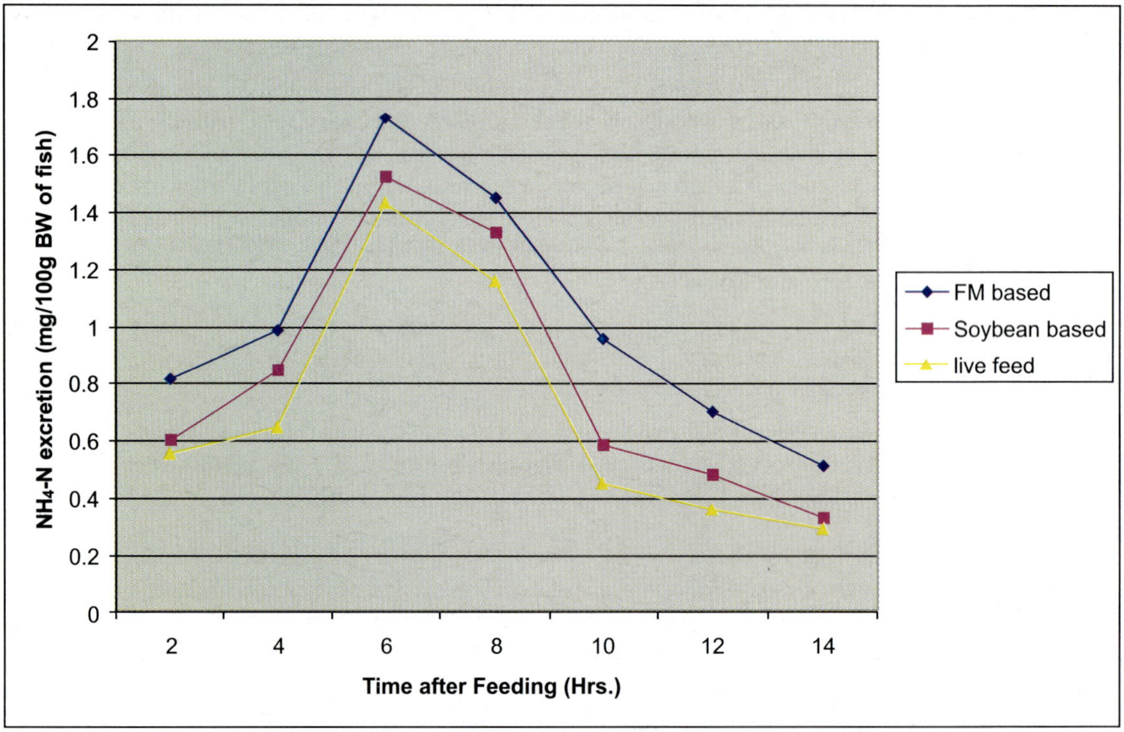

Figure 10.2: Diurnal Excretory Pattern of Ortho-Phosphate (O-PO_4) in *Clarias batrachus* Fry Fed on Live Feed and Supplementary Feeds (FM based and soybean based)

Table 10.5: Comparative Growth Performance of Catfish *Clarias batrachus* Fry Fed on Live Feed, Supplementary Feeds (FM based and soybean based) Under Laboratory Conditions

Sl.No.	*Type of Feed*	*Stocking Density*	*Survival Percent-age*	*Weight Gain (g)*	*Length Gain (cm)*	*Growth Per cent Gain in Body Weight*	*Growth/ Day in Per cent Body Weight*	*Specific Growth Rate (SGR) (g/day)*	*Specific Growth Rate (SGR) (cm/day)*
1.	Live food (zooplanktons)	20	97	3.15± 0.005	2.19± 0.003	105.75± 0.003	1.148± 0.003	1.45±. 0003	1.25± 0.003
2.	Soybean Based	20	95	2.84± 0.003	2.09± 0.004	99.98± 0.005	1.116± 0.001	1.37± 0.003	1.15± 0.005
3.	FM based	20	91	2.40± 0.005	1.95± 0.005	97.54± 0.003	1.012± 0.000	1.25± 0.002	1.04± 0.003

Experimental period of 70 days.

All values are mean±S.E. of means of 3 observations.

In the present investigation, best growth of *Clarias batrachus* fry was observed in zooplankton fed group indicating its importance. Fry fed with zooplankton showed a mean individual growth increment of 3.15g. and a high level of activity in 70 days. Our findings for zooplankton are in line with those of Geigher (1983 a,b); and Dabrowski (1984) Hoogendoorn (1980) and Hossain *et al.* (2006). Nayak *et al.* (2003) also reported inferior growth for *Clarias* fry using dry food (trout starter meal) alone in comparison to live feed. Growth of *Clarias lajera* (Hoogendoorn, 1980) and *Clarias gariepinus* (Polling *et al.*, 1988) using live food corresponded well with our finding.

Conclusion

Our present results strongly support the use of live food in the early growth phase of *C. batrachus*. Growth performance and activity of fry fed on supplementary feeds (FM based and soybean based) was not good in comparison to the live food. The poor quality of growth as well as less activity observed in the fry fed with supplementary feeds may be due to their inadequate acceptability at early growth phase. Low digestibility of feed might also caused such type of inferior growth. Although our present results strongly support the use of zooplankton as live food in the early growth phase of *C. batrachus* as they resulted in better growth and also excrete less ammonia and phosphorous in the water, while supplementary feeds give poor growth and very fluctuating survival rates. Further among supplementary feeds, soybean based feed give better results and excrete less ammonia and phosphorous rather than FM based feed showing plant based feed is best than animal based feed but live feed (zooplankton) is superior feed than supplementary (dry) feed. Therefore, the use of live feed not only resulted in better growth of fish fry *C. batrachus* but also less pollute the aquaria water as fry fed on live feed excrete less ammonia and phosphorous and thus alleviate pollution problems in the intensive aquacultural systems.

Acknowledgement

The authors acknowledge funding received under the scheme "Women Scientist Scholarship Scheme for Societal Programmes (WOS-B), Department of Science and Technology, Government of India" for carrying out this research. Chaudhary Charan Singh Haryana Agricultural University is thankfully acknowledged for providing necessary facilities.

References

AOAC (Association of Official Analytical Chemists), 1995. *Official Methods of Analysis*. Assoc. Off. Anal. Chem. Washington, Sc, USA.

APHA (American Public Health Association), 1998. *Standard Methods for the Examination of Water and Wastewater*, 20th edn. APHA, AWWA, EPFC, New York.

Dabrowski, K., 1984. Influence of Initial weight during the change from live to compound feed on the survival and growth of four cyprinids. *Aquaculture*, 40: 27–40.

Garg, S.K., 2003. *Recent Methods in Fish Nutrition and Biotechnology: Practical Manual*. Department of Zoology and Aquaculture, CCS HAU, Hisar, pp. 1–183.

Garg, S.K., Kalla, Alok and Bhatnagar, Anita, 2002. Evaluation of raw and hydrothermically processed leguminous seeds as supplementary feed for the growth of two Indian Major carp species. *Aquaculture Res.*, 33: 151–163.

Geiger, J.G., 1983a. Zooplankton production an manipulation in striped bass rearing ponds. *Aquaculture*, 35: 331–351.

Geiger, J.G., 1983b. A review of pond zooplankton production and fertilization for the culture of larval and fingerling stripped bass. *Aquaculture*, 35: 353–369.

Hoogendoorn, H., 1980. Controlled propagation of the African catfish, *Clarias lazera* (C and V). III, Feeding and growth of fry. *Aquaculture*, 21: 233–241.

Hossain, Quazizahangir, Hossain, M. Altaf and Parween, Selina, 2006. Artificial breeding and nursery practices of *Clarias batrachus* (Linnaeus,1758). *Scientific World*, 4(4): 32–37.

Jindal, M., Garg, S.K. and Yadava, N.K., 2007a. Effect of replacement of fishmeal with dietary protein sources of plant origin on the growth performance and nutrient retention in the fingerlings of *Channa punctatus* (Bloch.) for sustainable aquaculture. *Pb. Univ. Res. J. (Sci.)*, 57 (in press).

Jindal, M., Garg, S.K., Yadava, N.K. and Gupta, R.K., 2007b. Effect of replacement of fishmeal with processed soybean on growth performance and nutrient retention in *Channa punctatus* (Bloch) fingerlings. In: *Livestock Research for Rural Development*, Volume 19, Article #165. Retrieved from http://www.cipav.org.co/lrrd/lrrd19/11/jind19165.htm.

Jindal, Meenakshi and Garg, S.K., 2005. Effect of replacement of fishmeal with defatted canola on growth performance and nutrient retention in the fingerlings of *Channa punctatus* (Bloch.). *Pb. Univ. Res. J. (Sci.)*, 5: 183–189.

Jindal, Meenakshi, 2008. Studies on protein requirements of catfish *Clarias batrachus* fingerlings for sustainable aquaculture. In: *National Seminar on Technical Advances in Environment Management and Applied Zoology*, Jan. 23–25, Department of Zoology, Kurukshetra University, Kurukshetra, pp. 59–62.

Kalla, A., Bhatnagar, A. and Garg, S.K., 2004. Further studies on protein requirements of growing Indian major carps under field conditions. *Asian Fisheries Sci.*, 17: 191–200.

Nayak, P.K., Mishra, J., Kumar, K., Sahoo, S., Satpathy, B.B. and Ayyappan, S., 2003. Live food for the early larval growth of catfish, *Heteropneustes fossilis* (Bloch). *Indian J. Fish*, 50(3): 333–338.

Polling, L., Schoonbee, H.J., Prinsloo, J.F. and Wild, A.J.B., 1988. The evaluation of live feed in the early larval growth of the sharptooth catfish, *Clarias gariapinus* (Burchell). *Water SA*, 14(1): 19–24.

Prinsloo, J.F. and Schoonbee, J.H., 1986. Comparison of the early larval growth rates of the Chinese silver carp, *Hypophthalmichthys nobilis* using live and artificial feed. *Water SA*, 12(4): 229–234.

Steffens, W., 1989. *Principles of Fish Nutrition*. Ellis Horwood, Chichester.

Uys, W. and Hecht, T., 1985. Evaluation and preparation of an optimal dry feed for the primary nursing of *Clarias gariapinus* larvae (Pisces: Claridae). *Aquaculture*, 47: 173–183.

Chapter 11

Culture of Indian Magur

☆ *M.M. Girkar, S.B. Satam and S.S. Todkari*

Introduction

The most important aquaculture species of Asian catfishes *Clarias batrachus* (family claridae) have consumer preference and their culture systems are yet to be established in many countries of Asia. Among the catfishes, *Clarias batrachus* is of utmost importance owing to its taste, medicinal and high market value. This fish popularly known as magur, is an air breathing fish and well adapted to adverse ecological conditions. They normally inhabit swamps, marshy and derelict waters. These water bodies are usually shallow with heavy silt of decaying vegetation and organic load with poor nutrient release. Besides, these water-bodies have low pH, oxygen and primary productivity. On the contrary they have high carbon dioxide, hydrogen sulphide, methane and free ammonia, and this type of adverse environment is quite insensitive to the above air breathing slow growing, hardy omnivorous fishes. CIFA has standardized the following technologies on the breeding, seed production, larval rearing and grow out system of this precious species.

The suitability of this species for pond culture is based on the following biological criteria (Hogendoorn, 1979):

1. It matures and is relatively easy to reproduce in captivity
2. It can grow fast and efficiently
3. It supports high population densities
4. It is hardy in nature and it can tolerate wide range of environmental conditions
5. To adapt to fresh and brackish water.

Distribution

Clarias gariepinus locally called magur, is an indigenous species in Africa where it is widely distributed. It naturally inhabits tropical swamps, lakes, rivers and floodplains some of which are subject to seasonal drying. In recent years the species has been introduced in Europe, Asia and South-America. The species easily adapts to environments, where the water temperature is higher than 20°C.

Identification Characteristics

1. Its body is elongated cylindrical and anguilliform shape.
2. Magur has a scaleless slimy skin, which is darkly pigmented in the dorsal and lateral parts of the body.
3. The head is flattened, highly ossified, the skull bones (above and on the sides) forming a casque.
4. The length of head is 30–35 per cent of body length.
5. Around the mouth, 4 pairs of barbels can be distinguished (nasal, maxillary, the longest and most mobile, outer mandibular and inner mandibular). The magur can move the maxillary barbels independently of its mouth. The barbels serve as tentacles. Close to the nasal barbels, two olfactory organs are located. Magur recognizes its prey mainly by touch and smell. This is of relevance during feeding at night and in highly turbid or muddy waters where visibility is less.
6. Mouth is wide able to feed on a variety of food items ranging from minute zooplankton to large fish. They are able to suck benthos from the bottom, tear pieces of cadavers with the small teeth on its jaws and to swallow prey such as a whole fish.

Foods and Feeding Habit

Micha, (1973) considered magur as an omnivorous fish with a high tendency for predation. Different kinds of food items were found by different authors in the stomach of the African magur captured from natural waters. The food items reported are aquatic and terrestrial insects, fish, molluscs, fruits, diatoms, arachnids, plant debris, seeds, detritus, bird eggs, young birds, droppings of poultry and zooplankton. In zooplankton-rich fish pond, African magur often join other carp species to graze zooplankton on the surface. Many authors concluded that the African magur is an omnivorous slow moving predatory fish which feeds on a wide variety of food items from zooplankton to fishes of half of its own length (Janssen 1987).

Culture Techniques

General Considerations

Various production conditions can be found in Bangladesh. It seems that magur can be adopted to these wide range of possibilities. However, intensity of production depends on the following parameters:

Pond Size

A pond of 0.75–1.5 ha is ideal for production of magur. Most of the small ponds around the farm are suitable for magur stocking after some modifications. But the large, more than 1 acre undrainable ponds are not recommended for stocking with African magur. For large scale production properly designed ponds should be constructed. Depth of pond for grow out should be within the range of 1.5–2.0 m.

Availability of Fingerlings

Year-round, multi-crop culture of African magur largely depends on availability of fingerlings throughout the year. Under the local conditions of Bangladesh, it is now possible to breed magur for at least eight months in a year.

Availability of Inputs

The input demand increases exponentially with the intensity of production. But in small scale culture where magur, is a small component in polyculture, the input cost is little. With the increase in intensity of production the input costs rise sharply.

Availability of Feed Ingredients

For African magur culture, the protein feed supply is the main limiting factor. One should find out first where the slaughterhouse by products or other cheap animal protein sources are available on a daily basis and in adequate quantities.

Marketing

The present market system (mainly in rural areas) is favourable to small scale operations. When the production of African magur is specialized and intensified, the market channel should be found out before stocking of fingerlings. For commercial production of magur, detailed market study should be undertaken on such operations.

Preparing Pond for Magur Production

Before deciding to construct a new pond for magur culture the followings should be checked:

1. The area should be free from flood free even at the highest flood level.
2. Water retention capacity of the soil is good.
3. Dig a 1–1.2 m hole on the proposed site and fill the hole with water. Next day fill again to make up the water loss by seepage. After 24 hours the water level and the wall should be checked. If the water level remain more or less same the soil is suitable for magur pond.
4. While making embankments, roots, leaves and other organic particles should be removed from the soil, and the soil should be compacted well. Local traditions and low initial input demands support the simple undrainable ponds.
5. The pond bottom should be prepared flat with slight slope towards the outlet or the usual pumping place. Grass should be planted on the top and the slopes of dykes to reduce erosion.

Eradication of Predator and Weed Fishes

Before stocking fingerlings in the pond it is necessary to eradicate the unwanted fish from the pond. Predator and weed fish tend to decrease yield considerably. Repeated netting may not be sufficient. It is, therefore, necessary to apply fish toxicants. Some of the efficient toxicants listed in Table 11.1 can be used.

Liming

Liming of a fish pond is highly recommended if the soil is not alkaline. Lime neutralizes soil acidity and creates a buffer system to prevent marked diurnal changes of the water from acidic to alkaline conditions. Apart from other advantages, this buffering action of calcium is the most important. Lime treatment for ponds should be done before initial manuring as in Table 11.2.

Table 11.1: Efficient Toxicants and their doses concentrations

Efficient Toxicants	*Concentration of Dose (ppm)*
Rotenone	2–3 ppm
Phostoxin	0.2 ppm (Toxicity lasts for around 10–15 days)
Tea seed cake	75–100 ppm. (The toxicity lasts for about 10–12 days)
Mahua oil cake	250 ppm. (The toxicity lasts for about 10–15 days)
Bleaching Powder	30 ppm. (Its toxicity lasts for about 7–8 days)

Table 11.2: Soil Type and Preferable Lime Dosages

Soil pH	*Soil Type*	*Quantity of lime ($CaCO_3$)*
4.0–4.9	Highly acidic	270 (Kg/ha)
5.0–6.4	Moderately acidic	140 (Kg/ha)
6.5–7.4	Near neutral	70 (Kg/ha)
7.5–8.4	Mildly alkaline	30 (Kg/ha)
8.5–9.5	Highly alkaline	No liming

Initial Fertilization

A well prepared zooplankton rich water is the best starter for magur fingerlings. The magur fingerlings try to supplement its higher protein requirement from natural sources. At stocking when the size is 2–3 cm, the bigger size zooplankton *e.g. Cladocera sp., Diaptomus* is the cheapest and best food source for fingerlings.

In different types of magur culture in earthen ponds considerable quantities of nutrient elements are removed from the pond ecosystem through fish production. Therefore the fertilizer requirements vary among different soil productivity levels and the intensity of production. Organic manuring besides being important as means of adding the nutrients is also equally important for improving the soil texture. During the shortage of organic manures, application of inorganic fertilizers is recommended.

Table 11.3: Manuers and their Suitable Dosages

Manures/Fertilizers	*Quantity (kg/ha)*
Chicken manure (dry) or	250
Cow dung	400
Urea	20
Triple/Single Super Phosphate	7

Induced Breeding Techniques

Usually this species attains maturity at the age of one year and breeds during June-August. During breeding season male and female magur are collected from brood stock pond, and kept separately in plastic tubs containing water. Males and females are distinguished by their secondary sexual characters. In females the abdomen is gravid, vent is reddish colored, genital papilla is round and button shaped. In male genital papilla is elongated and pointed. The brood fish of 100-150 g size can be used for successful spawning. For the purpose, they should be well fed with feed containing 30 per cent protein diet daily, at least 3 months prior to breeding season. The females are either induced through hormone therapy using carp pituitary extract (30-40 mg/kg), HCG (4000 IU/kg) and Ovaprim or Ovatide (0.6-0.8 ml/kg) or through environmental manipulation in a controlled system for spontaneous spawning. The injected fishes are to be stripped after 14-17 hr for releasing eggs and fertilized artificially with sperm suspension made with mascerating the testes of the male fish. Shining and brown coloured beads like eggs are considered as good eggs, while white coloured eggs are of bad quality. A 150 gm female lays around 7,000-9,000 eggs.

The flow-through hatchery consist a row of small plastic tubs of 12 cm diameter, 6 cm height placed on a cement platform and are provided with flow-through water system. The water supply is provided from an over head tank through a common pipe to all the tubs with individual control taps. Each tub is having the provision of an outlet at a height of about 4 cm. The fertilized eggs are uniformly distributed in the incubation tubs.

For large scale hatching, an improvised hatchery system has also been developed, which consists of a circular tank having 2 m diameter with inlets at a height of 15 cm at an angle of 45°. A feeble inflow of water is maintained. The eggs are uniformly spread for hatching. The system can accommodate 1 lakh fertilized eggs at a single operation with 60-80 per cent hatching rate. After hatching, the larvae are collected by siphoning from the bottom of the hatching tank.

Ideal temperature for hatching is between 27-30°C. Hatching takes place within 24-26 hours. The tubs from the flow-through hatchery are washed properly to make the larvae free from egg shell. The magur larvae with yolk sac measures about 4-5 mm in length.

Larval Rearing

The larvae are reared in indoor rearing tanks. There is no necessity to provide feed during first three days as yolksac in larvae serves as the stored feed. After yolksac absorption, the larvae are fed with either live plankton or *Artemia* nauplii. It is important to provide good environment to the larvae. For the purpose the indoor rearing tanks are provided with continuous aeration and water exchange facilities. There is a chance of mortality and poor growth of larvae due to poor environment and high stocking density during in-door rearing phase. A stocking density of 2000–3000 nos/m^2 is considered to be optimum for better growth and survival in indoor condition. The larvae grow to 10–20 mm or 30–40 mg fry during 12–14 days of rearing. After a maximum of 14 days rearing in the indoor, they should be transferred to out-door rearing tanks for fingerling production.

The advanced fry are further reared in cemented cisterns or earthen ponds for fingerling production. Generally the advanced fry reared in pond condition do not show good survival due to natural mortality or predation as in this stage the fish does not have much capacity to escape from predators. Therefore, small cemented tanks of 10-20 m^2 size are required for better survival and easy management. The tanks are prepared with soil base and manured like carp nursery ponds. The tanks are inoculated with plankton and the advanced fry are stocked after 6-7 days of preparation. The fry grows to a size of 0.8-1 g during 30 days of rearing when fed with laboratory made feed containing 30 per cent protein and at a density of 200-300 nos/m^2.

Stocking of Pond

The fry and fingerling can be transported from nurseries when the pond is ready for stocking. The fry is ready for stocking when the arborescent organs are developed. It is completed after about two weeks of age. It is easy to identify by the special behavior of air gulping. The small magur fry vertically swims up to the surface to breathe air and quickly goes down. Size of fingerlings should be equal. Size differences in magur naturally occur from the beginning of rearing. Fingerlings of equal size should be sorted out from the stock. The fingerlings may be kept in a net for a short time, the mesh size of which will allow the small ones swim out. The bigger size fingerlings should be picked up by hand during counting. The initial size differences will be even more during the grow-out period and that difference creates problems like cannibalism. Table 11.4 should be help in identifying the healthy from the non-healthy fingerlings:

Table 11.4: Identification of Major Fingerlings

Sl.No.	*Healthy magur fingerlings*	*Non-Healthy magur fingerlings*
1.	Quick moving and gathering on the bottom of the net	Fingerlings hang vertically on the surface with weak response to disturbances
2.	Fast swimming up and down when gulping air	Slugish swimming up and down when gulping air
3.	Unbroken barbels and fins	Barbels and fins are broken
4.	Skin is uniformly dark-grey colour covered with mucus.	Skin is dull dark-grey colour covered with no mucus.

For short distance transportation magur fry can be transported in plastic buckets with led. For long distance transportation, plastic bags with oxygen should be used. For this, the fingerlings should be held in a hapa for 12 hours to get their guts cleared. Otherwise, decaying faeces in the water inside the plastic bag produces ammonia which may cause high mortality during long distance transportation. Before releasing in the pond, the fingerlings should be acclimatized to the temperature of the pond water. This can be done by slowly adding pond water into the water containing the fingerlings inside the plastic bag.

The stocking rate of magur and the supplementary carp species will vary according to (*i*) the availability of feed for magur and (*ii*) the pond conditions, especially on the quantity and the quantity of natural feed present in the pond. However, the following stocking rate is recommended for semi-intensive magur dominated polyculture system.

Grow Out Culture

The earthen ponds or stone pitched ponds or cemented tanks are suitable for grow-out culture of magur. Generally high density of 50,000-70,000/ha is recommended for culture of this species. Bigger sized fish (5-10 g) shows good survival and growth during culture. The fishes are fed at the rate of 3-5 per cent of their body weight with pelleted feed in the feeding basket placed in different places of the pond.

Since they are air breather, they normally come up to the water surface for atmospheric oxygen. This kind of habit attracts birds for predation. Therefore, it is required to cover the ponds with net to protect the fishes. The fishes attain a marketable size of 100-120 g during a period of 7-8 months. Harvesting of magur is done by dewatering the pond completely and picking them manually from the culture ponds. Productions of 3-4 tonnes are achieved from one hectare of water area.

Feed and Feeding

In semi-intensive production level the growth of magur is highly dependent on the quality and quantity of supplementary feed. As the stocking rate in the pond increases the feed requirement also increases. Protein is the most expensive component of the magur supplementary feed and is the most important nutrient in the diet. Protein sources should be first identified from the ingredients available within reasonable distance. In semi-intensive pond culture most of the protein demand must be met through the supplementary feed. Smaller part is gathered by the fish from the well prepared pond water. Only a small part of the protein requirement may be met from the natural feed available in the pond. The basic level of nutrient requirement of African magur is given in Table 11.5.

Table 11.5: Nutrient Requirement of African Magur (Modified after Janssen 1987)

Nutrients	*Per cent of Dry Matter*
Protein	30–35
Digestible energy	2,500–3,500 Kcal/kg
Ca	0.5–1.8
P (available)	0.5–1.0

Common Diseases in Magur

In well managed culture conditions African magur is found to be more or less resistant to fish pathogens *viz.* virus, bacteria and parasites. Poor water quality, infected feed and rough handling can make the fish weaker and make them susceptible to diseases.

Stressed or infected fish can be easily recognized by some abnormal behaviours, such as poor appetite and abnormal swimming behaviour, (staying in vertical position on the pond surface). In addition, some clinical symptoms such as mutilated barbels or fins, white or red-brown spots on the skin, pop-eyes etc. could be observed. The health conditions of magur should be monitored daily, particularly during feeding when magur frequently come to the surface water. For diagnosing bacterial, fungal and parasitic diseases, squash preparations of the skin, the gill filaments, intestines etc. have to be made and examined.

Bacterial Infections

Columnaris Disease

Caused by *Flexibacter columnaris*

Symptoms The fish remain in vertical position at the water surface. Big white spots like lesions on the body without mucus. Fins are broken.

Treatment Antibiotics such as Chloramphenicol, Terramycin or Oxytetracycline can be added to the feed. Dose: 5 to 7.5 g/100 kg fish/day for 5–10 days.

Hemorrhagic Septicaemia

Caused by *Aeromonas hydrophila, Pseudomonas fluorescents,* etc.

Symptoms Shallow ulcerations, hemorrhages and in severe cases, swollen abdomen. Internally the body cavity is filled with opaque fluid. Pale liver and sometimes haemorrhages over swim bladder.

Treatment Water change and check of feed quality. Terramycin (oxytetracycline) with feed 7.5 g/100 kg body weight/day for 10–12 days. Furazolidone 5–7.5 g/100 kg body weight/day for 2–3 weeks. Pond treatment with 3–5 ppm of potassium permanganate.

Fungal Disease: ***Saprolegniasis***

Caused by *Saprolegnia* spp.

Symptoms Ulceration on the skin, fin erosion, exposure of muscles and jaw bones and in some cases tufts of minute white hair like out growths may occur in the affected parts. The disease develops mainly in winter period.

Treatment Dip treatment in 3 per cent common salt solution for 20–30 minutes or bath in 0.1 ppm Malachite Green.

Diseases Due to Malnutrition and Environmental Stress

Symptoms Pop-eyes, soft skull and sometimes deformed caudal fins are present. In a later stage of the disease a gradual destruction of the arborescent organs occurs. Knocking on the skull of affected fish produces hallow sound. Delayed calcification and finally breaking of skull. The disease is particularly prevalent in magur larger than 10 cm.

Treatment Bad pond water conditions such as polluted water, and bad quality feed. The pond water should be exchanged and the flow rate of the water should be increased when the first symptoms of the disease appear. The supply of feed should be stopped for a few days and preferably be replaced by fresh and vitamin C added feed.

Harvesting and Marketing

Partial harvest of magur may start after 50–60 days of culture (except in carp polyculture), when some specimens reach marketable size (above 200 g).

Magur is a clever, fast moving fish. Proper nets are needed for catching magur effectively from the pond. A lot of magur usually escapes through under the bottom line of a net.

Harvesting should be done gently and quickly by seine net preferably in morning when the temperature is cooler. During harvesting, marketable fish should be sorted out first and then small size fish should be released back to pond. In polyculture or magur dominated polyculture the carp species should be sorted out first. The total operation should be done as quickly as possible so that the fishes returned to pond are not stressed.

Harvesting and marketing of fish in rural areas has to be adjusted in accordance with the market days–usually twice a week. Marketing of small quantity of fishes in batches would ensure better price in local rural markets.

References

Bruton, M.N., 1976. On the size reached by *Clarias gariepinus*. *J. Limnol. Soc., S. Africa*, 2: 57–58.

Haniffa, M.A., Jesu Arockia Raj, A. and Arul Mozhi Varma, T., 2001. Optimum rearing conditions for successful artificial propagation of catfish. NBFGR-NATP Publication No.3. Captive breeding of aquaculture and fish germplasm conservation. Paper No. 4.

Hogendoorn, H., 1979. Controlled propagation of African catfish *Clarias Tazera* (C and V). I. Reproductive biology and field experiment. *Aquaculture*, 17(4): 323–333.

Islam, M.N., Rahman, S.M., Hossain, Q.Z., Ahsan, M.N. and Asaduzzaman, S.M., 2004. Effect of live and formulated diets on growth and survival of *Clarias batrachus* larvae. In: *Proc. 14th Biennial Nat. Conf.*, The Zoological Society of Bangladesh, Dhaka, 26–27, February, pp. 8.

Janseen, J., 1987. Hatchery management of the African Clariid Catfish *Clarias gariepinus* (Burechell, 1822). In: *Selected Aspects of Warmwater Fish Culture*, (Eds.) A. Coche, and D. Edwards. FAO/ UNDP, Rome, 1989, 181pp.

Kumar, D. *Manual of Fish Culture in Undrainable Ponds*. FAO Publication (in press).

Micha, J.C., 1973. Eludedis populations piscicoles de I. vbangui et tentatives de selection et d'adaptators de quelques especes a l'etany de piscicultue-Nogent-san-Marine, Centre Technique Forestier Tropical, 110p.

Morris, Joseph E. *Best Management Practices for Channel Catfish Culture*. Department of Animal Ecology 124 Science II, Iowa State UniversityAmes, IA 50011–3221.

Morris, Joseph E., 1993. *Pond Culture of Channel Catfish in the North Central Region*. North Central Regional Aquaculture Centre.

Munshi, J.S.D., 1996. *Ecology of Heteropneustes fossilis: An Air-Breathing Catfish of South-east Asia*. Narendra Publishing House, Delhi, pp. 174.

NACA, 1989. *Integrated Fish Farming in China*. NACA Technical Manual 7. A World Food Day Publication of the Network of Aquaculture Centres in Asia and the Pacific, Bangkok, Thailand, 278 pp.

New, M.B., 1987. *Feed and Feeding of Fish and Shrimp*. ADCP/REP/87/27, FAO/UNDP, 275pp.

Quazizahangir Hossain, M. Altaf Hossain and Parween, Selina, 2006. Artificial breeding and nursery practices of *Clarias batrachus* (Linnaeus, 1758). *Scientific World*, 4: 82–87.

Raghavan, Rajeev, 2006. Potential culture species in Southwestern India. *Global Aquaculture Advocate*, pp. 68–69.

Rao, G.R., Tripathi, S.D. and Sahu, A.K., 1994. *Breeding and Seed Production of the Asian Catfish Clarias batrachus* (Lin). Central Institute of Freshwater Aquaculture, Barrackpore, pp. 47.

Saha, M.R., 1996. Effects of various doses of ovaprim for breeding of Clarias spp. in Tripura. *Journal of the Inland Fisheries Society of India*, 28(2): 75–84.

Saha, M.R., Mollah, M.F.A. and Roy, P.K., 1998. Growth and survival of *Clarias batrachus* (Linn.) larvae fed on formulated diets. *Bangladesh Journal of Fisheries Research*, 2(2): 151–158.

Saha, M.R., Mollah, M.F.A. and Roy, P.K., 1998. Rearing of *Clarias batrachus* (Linn.) larvae with formulated diets. *Bangladesh Journal of Fisheries Research*, 2(1): 41–46.

Sahoo, S.K., Giri, S.S., Chandra, S. and Sahu, A.K., 2007. Spawning performance and egg quality of Asian catfish *Clarias batrachus* (Linn.) at various doses of human chorionic gonadotropin (HCG) injection and latency periods during spawning induction. *Aquaculture*, 266(1–4): 289–292.

Sambhu, C., 2004. African catfish, *Clarias gariepinus* (Burchell, 1822): An ideal candidate for biowaste management. *Indian J. Exp. Biol.*, 42(12): 1226–1229.

Singh, A.K. and Mishra, A., 2001. Environmental issue of exotic catfish culture in Uttar Pradesh. *J. Environ. Biol.*, 22(3): 205–208.

Tripathi, Satyendra Datt, 1996. Present status of breeding and culture of catfishes in South Asia. *Aquat. Living Resour.*, 9: 219–228.

Vijayakumar, C., Sridhar, S. and Haniffa, M.A., 1998. Low cost breeding and hatching techniques for the catfish *Heteropnuestes fossilius* for small-scale farmers. *NAGA*, 21(4): 15–17.

Viveen, W.J.A.R., Richter, C.J.J., Van Oordt, P.G.W.J., Janssen, J.A.L. and Huisman, E.A. 1986. *Practical Manual for the Culture of the African Catfish (Clarias gariepinus).* University of Wageningen, Netherlands, 121p.

Yasmin, A., Mollah, M.F.A. and Haylor, G.S., 1998. Rearing of Catfish (*Clarias batrachus*, Lin.) Larvae with live and prepared feeds. *Bangladesh Journal of Fisheries Research*, 2(2): 145–150.

Chapter 12

Influence of Heavy Metals on Abundance of Cyanophyceae Members in Three Spring-Fed Lake in Kempty, Dehradun

☆ *P.K. Bharti, D.S. Malik and Rashmi Yadav*

ABSTRACT

The present chapter deals with the water quality status of three lakes depended on Kempty fall by the assessment of physico-chemical parameters and planktonic diversity and the relationship of planktonic population and heavy metals. Temperature, pH, DO, free CO_2, BOD, COD, hardness, chloride, sodium, potassium, calcium, magnesium, zinc, iron, manganese and cobalt were analysed for every sampling station. Physico-chemical parameters and some heavy metals were observed with slightly variation in all three lakes. In the study Zn and Fe were found as a regulatory key for the population of cyanophyceae members.

Introduction

Metals that are naturally introduced into the waterbody come primarily from such sources as rock weathering, soil erosion, or the dissolution of water-soluble salts. Naturally occurring metals move through aquatic environments independently of human activities, usually without any detrimental effects.

Humans consume metallic elements through both water and food. Some metals such as sodium, potassium, magnesium, calcium, and iron are found in living tissue and are essential to human life-biological anomalies arise when they are depleted or removed. Probably less well known is that currently no less than six other heavy metals including molybdenum, manganese, cobalt, copper, and zinc, have been linked to human growth, development, achievement, and reproduction (Vahrenkamp, 1979; Friberg *et al.*, 1979). Even these metals, however, can become toxic or aesthetically undesirable when their concentrations are too great. Several heavy metals, like cadmium, lead, and mercury, are highly toxic at relatively low concentrations, can accumulate in body tissues over long periods of time, and are nonessential for human health.

The toxic heavy metals entering the ecosystem may lead to geo-accumulation, bioaccumulation and biomagnifications. Heavy metals like Fe, Cu, Zn, Ni and other trace elements are important for proper functioning of biological systems and their deficiency or excess could lead to a number of disorders. Food chain contamination by heavy metals has become a burning issue in recent years because of their potential accumulation in biosystems through contaminated water, soil and air. Therefore, a better understanding of heavy metal sources, their accumulation in the soil and the effect of their presence in water and soil on plant systems seem to be particularly important issues of present-day research on risk assessments (Lokeshwari and Chandrappa, 2006).

Water is an important substance required by all living organisms and for all anthropogenic activities (Dagaonkar and Saksena, 1992). The problem of environmental pollution due to toxic metals has begun to cause concern now in most major metropolitan cities (Bharti, 2007). Most of our water resources are gradually becoming polluted due to the addition of foreign materials from the surroundings. These include organic matter origin from plant and animal, land surface washing, and industrial and sewage effluents. The lakes have a complex and fragile ecosystem, as they do not have self-cleaning ability and therefore readily accumulate pollutants. In some tourist hilly regions, there are not any more industrial pollution, but tourist's anthropogenic activities may alter the water quality and the water quality of lakes of some tourist spot may affect the biological diversity of that water body.

Study Area

Kempty village is situated at latitude of 30° 27′ N and longitude of 78° 06′ E, 13 km far from Mussoorie city. A natural spring known as Kempty fall is the main attraction of tourists since a prolonged time, which maintain the water label of nearest three lakes. Due to some tourist activities, the spring water may affect and goes to lakes. Some anthropogenic activities may change the ratio of abiotic and biotic components of an ecological system.

Methodology

To assess the water quality of three lakes, Upper lake Kempty was selected as the first sampling station S1 and other assistant lakes were selected as S2 and S3. Water samples were collected in plastic jericanes and analyzed according to APHA (1995) and Trivedi and Goel (1984).

Results and Discussion

Various physico-chemical and planktonic parameters have been depicted in Tables 12.1 and 12.3. The mean value water temperature of lakes was recorded 20.83 due to the cool climate. Physico-chemical characteristics of all lakes were found to be low, under permissible limit. Water pH varies between 6.8-7.2 and likely neutral on pH scale. Dissolved oxygen was found more in first upper lake due to proper aeration during the water falls from spring. Oppositely, the values of BOD, COD and free carbon dioxide were decreases according to oxygen content in lake water respectively. Dissolved oxygen has negative correlation among free carbon dioixe, BOD, COD and temperature (Malik and

Bharti, 2005). Less chloride was found 14 mg/l in first Lake S1 and reach 19 mg/l in third lake due to various anthropogenic interferences. Sodium and potassium were found to be low in comparison of other hill stream like Sahastradhara. Heavy metals in three lakes were detected in very low concentration while upper lake has water free with some metallic contents due to the absence of industrial pollution.

Table 12.1: Physico-chemical Characteristics of Three Lakes at Kempty (mg/l)

Sl.No.	*Parameter*	*S1 (Lake1)*	*S2 (Lake 2)*	*S3 (Lake 3)*	*Mean*
1,	Temperature	19.5	20	20.5	20
2.	pH	7.2	7.5	6.9	7.2
3.	TS	270	280	350	283.33
4.	DO	9.5	7	6.9	7.8
5.	BOD	5.2	6.1	6.5	5.93
6.	COD	19	23	26	22.66
7.	Free CO_2	2.8	2.9	3.2	2.96
8.	Hardness	170	186	175	177
9.	Chloride	19	22	20	20.33
10.	Sodium	26	30	32	29.33
11.	Potassium	4.2	4.8	5.1	4.7
12.	Calcium	80	92	95	89
13.	Magnesium	56	68	70	64.66

Table 12.2: Heavy Metals in Three Lakes at Kempty (mg/l)

Sl.No.	*Parameter*	*S1 (Lake1)*	*S2 (Lake 2)*	*S3 (Lake 3)*	*Mean*
1.	Zinc	0.030	0.035	0.032	0.032
2.	Iron	0.655	0.679	0.668	0.667
3.	Manganese	0.125	0.132	0.142	0.133
4.	Cobalt	0.040	0.035	0.041	0.0386

Table 12.3: Planktonic Diversity in Three Lakes at Kempty (ind./l)

Sl.No.	*Parameter*	*S1 (Lake1)*	*S2 (Lake 2)*	*S3 (Lake 3)*	*Mean*
1.	Chlorophyceae	1110	1335	1543	1329.33
2.	Bacilliriophyceae	460	485	515	486.66
3.	Cyanophyceae	160	120	207	162.33
4.	Total phytoplankton	1730	1940	2265	1978.33
5.	Zooplankton	235	330	245	248.0
6.	Total plankton	1965	2270	2510	2248.33

Planktonic diversity generally has the abundance of phytoplankton of chlorophyceae family in S1 and S3 lakes. Both of the lakes having greenish colour of water but S2 lake has an abundance of

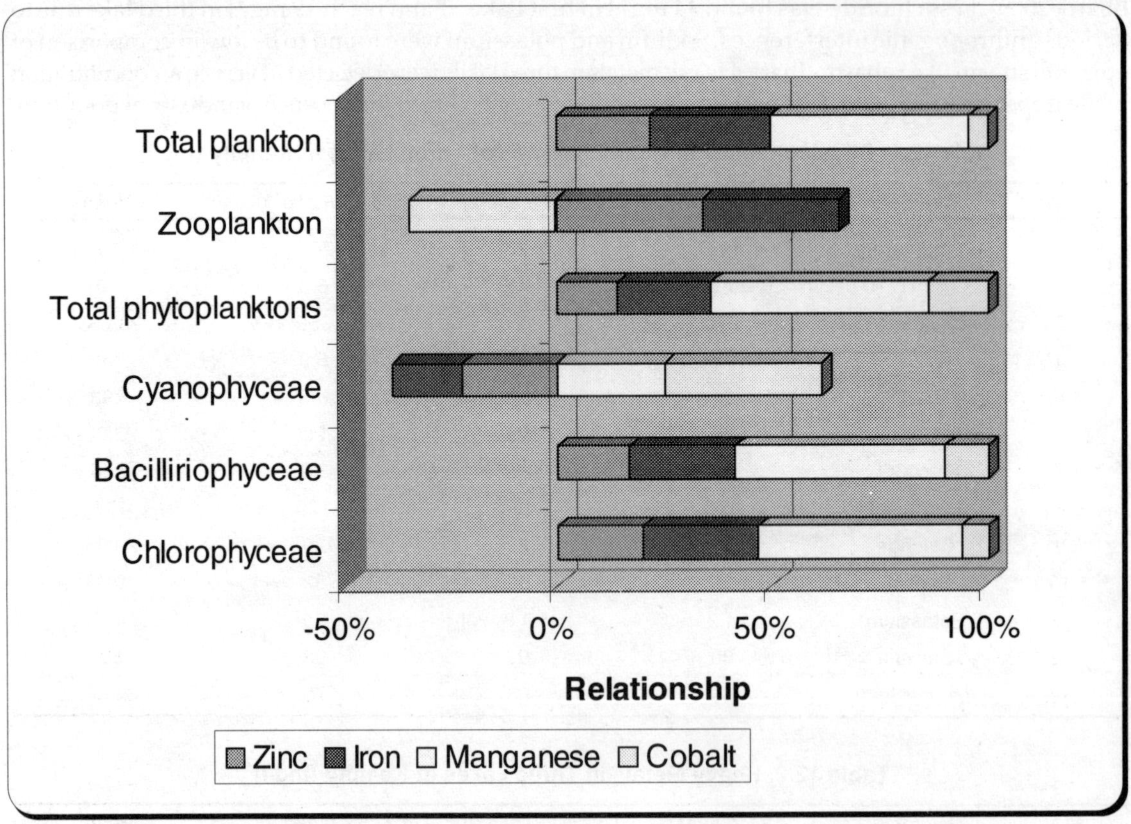

Figure 12.1: Relationship Between Planktonic Population and Heavy Metals

phytoplankton of cyanophyceae family so it seems bluish from Kempty fall view point (3 km behind Kempty fall). The number of phytoplankton of cyanophyceae was found maximum 273 individuals per liter due to its relationship with heavy metals, while rest two lakes have the abundance of chlorophyceae. Similar trend of planktonic distribution is also supported by Pande and Mishra, (2000). It was concluded in the present study that Zn and Fe were found having more influence on the population of cyanophyceae members and found to be an important factor for the distribution of cyanophyceae plankton in the three lakes at Kempty.

References

APHA, 1995. *Standard Methods for Examination of Water and Wastewater*, 19th Edn. American Public Health Association Inc., New York, pp. 1170.

Bharti, P.K., 2007. Where is environmental science is going in India? *Current Science*, 92(4): 414.

Dagaonkar, A. and Saksena, D.N., 1992. Physico-chemical and biological characterization on tample tank, Kaila Sagar, Gwalior. *J. Hydrobiology*, 8: 11–19.

Friberg, L., Nordberg, G.F., and Vouk, V.B. (Eds.) 1979. *Handbook on the Toxicology of Metals*. Elsevier/North–Holland Biomedical Press, Amsterdam, 709 p.

Lokeshwari, H. and Chandrappa, G.T., 2006. Impact of heavy metal contamination of Bellandur Lake on soil and cultivated vegetation. *Current Science*, 91(9): 622–627.

Malik, D.S. and Bharti, P.K., 2005. Nutrient dynamics in Rhithron zone of Shivalik Himalayan stream Sahastradhara, Dehradun (Uttaranchal). *Env. Cons. J.*, 6(2): 63–68.

Pande, R.K. and Mishra, A., 2000. Water quality study of freshwaters of Dehradun (Sahastradhara stream and Mussoorie lake). *Aquacult.*, 1(1): 57–62.

Trivedi, R.K. and Goel, P.K., 1984. *Chemical and Biological Methods for Water Pollution Studies.* Environmental Publication, Karad, pp. 1–251.

Vahrenkamp, H., 1979. Metalle in Lebensprozessen: Chemie in Unserer Zeit, 7: 97–105.

Chapter 13

Aquarium Keeping and Maintenance of Marine Ornamental Invertebrates

☆ *K. Balaji, G. Thirumaran, R. Arumugam, K.P. Kumaraguruvasagam and P. Anantharaman*

ABSTRACT

Marine ornamental invertebrate's aquarium behaviours were studied in the laboratory, three different types of tanks such as gravel, sand and *Uca* crab tanks as model habitat were prepared for different types of ornamental invertebrates on choice based. *Uca* crab and land hermit crab are ideal for the Uca crab type aquarium. Several diets (Pellet feed and live feed) have been tested in different form of invertebrate species, the feeding behaviour and the suitability of the food preferences to be species dependent. Plankton and newly hatched *Artemia* sp. are suitable for jellyfishes and filter feeders like *Lepas anserifera* and *Perna viridis*. Remaining invertebrates take minced and squashed fish and shrimp meat. Most of the study animals show the maximum longevity period of about more than hundred days and jellyfish *Chrysaora quinquecirrha* and *Rhizostoma* sp. shows the minimum longevity period of about one week. Sponge crab *Dromia dehanni* shows the maximum survival rate of about 90 per cent. *Chrysaora quinquecirrha* and *Rhizostoma* sp. shows the minimum survival rate of 0 per cent. Most of the study animals were compatible with both fishes and invertebrates, apart from angel crab *Charybdis feriata*. Knowledge about aquarium behaviour, feeding, food preference, longevity, survival and compatibility with fishes and among invertebrates are needed to keep and maintain the ornamental invertebrates in the aquarium. These findings will surely enable to keep and maintain the ornamental invertebrates in aquarium.

Introduction

Among ornamental fishes, marine fishes are attractive and colourful one, and its taming requires skill. Like that, ornamental invertebrates are very attractive colourful and interesting to keep in aquarium. For this, many different skills are needed to keep the invertebrates in aquarium. Many of the invertebrates are normally sedentary; some are slow moving from their coral and rock habitats. In the sense of aquarium keeping and maintenance of ornamental invertebrates, one should look to the base covering of the tank, feeding, compatibility, longevity and survival rate.

Base covering or aquascaping of aquarium should have the following purposes: to provide many functional benefits for the tank species, to provide camouflage for the fishes and invertebrates, to provide a growing medium of the aquarium plants, rocks and gravels provide comfort (for attachment, substratum) and security (as a hiding place) for the fishes and invertebrates, it may be used as spawning sites for some captive breeders, escape by some species, as a filtration system and that resembles the environment of the species from the wild, which gives more attractive aquarium. Thus, aquascaping of the tank related with the animals were from habitat nature to the tank and life style of the species, which is a challenge and very vital part of the aquarium keeping. The specific aquascaping for different invertebrates are dealt within detail in this chapter.

Invertebrates have no single feeding methods. Feeding methods vary from being scavengers, such as shrimps, hermit crabs and sea stars; filter feeders, such as sea whips, sea fans and clams; some are grazers such as many molluscs, sea stars and sea urchins; active predators such as octopus and squid; many sea anemones are propitious feeders. Depending on the types of invertebrates need to provide feeds. So we must give a wide variety of foods for different invertebrates. Liquid suspension foods for filter feeders and particulate foods for the scavengers.

Feeding methods vary in many invertebrates indeed they need to be fed individually. The mode of feeding is important. For sea anemones, shrimps and crabs hand feeding is essential. Liquidized fish meat or planktonic feeds in the aquarium must be provided for the filter feeders. Uneaten food will be taken by the scavengers such as crabs, hermit crabs, shrimps, starfishes, sea cucumbers and sea urchins. It is not essential to feed invertebrates as frequently as fish. For example, sea anemones could be fed once in two days. So, to keep invertebrates you will have to make special provision for feeding them. The specific food and feeding methods of many invertebrates are dealt with in detail in this study.

Compatibility of the tank mates provides good tank and healthier tank; otherwise some members will eat, or peck or kill the tank mates. Careful selection or sufficient care of compatible tank mates should avoid such problems. With selection of compatibility, crowd of the tank mates also important of the normal survival of the tank animals. Many invertebrates associate with some fishes, such as corals, anemones, starfish, tubeworms, shrimps, sea urchins and many other animals. Alternatively, we can set up a completely independent invertebrate collection, as a fish only tank.

To keep invertebrates successfully in an aquarium one should need skills like selection of species, aquarium behaviour, feeding methods, food preference, longevity, survival and compatibility with fishes and among invertebrates. Because many are normally sedentary, hardly moving from their habitats, special care is needed to maintain the invertebrates in captive. The present investigation also experienced to keeping the marine ornamental invertebrates in low cost expenses and is kept with the basic principles and equipments.

Materials and Methods

Rearing Facilities

For the purpose of aquarium keeping, certain numbers of live and healthy animals from each species were transported to the laboratory in an aerated condition. The animals were maintained throughout the study period in glass aquaria and plastic troughs of 54 and 60 liters water holding capacity respectively. Filtered and UV treated seawater (33-35 per cent) was used. Continuous aeration was provided with the help of diaphragm type aerators. Gravels size of around 3 mm, decorative stones, rocks, different size of gastropod's dead shells, arch shaped tile as hiding place and clear sand were used. Bio-filter was used under the gravel, sand type and Uca crab aquarium. About 25 per cent of water exchange was given every day. Water quality parameters like pH, salinity, and temperature of the aquaria were regularly monitored. The water temperature (°C) was measured using a standard Celsius thermometer with the accuracy of ± 0.5°C. Salinity (per cent) was estimated with the help of an E-2 model salinometre. The dissolved oxygen concentration was estimated using the Winkler's method as described by Strickland and Parsons (1972). The pH was measured using an Elico-model LI 10 pH pen.

Scoop net was used to collect the larvae of freshwater prawn, *Acetes* sp., and fish larvae during high tide time. Plankton was collected from Vellar estuary using plankton net (No. 30, mesh size: 48 μm) having a mouth area of 0.35 diameter. Collected plankton soup was added to the aquarium water for filter feeders. For feeding purposes glass Petri plates, glass beakers, aerator, air stones, pointed forceps, needle was used. With the help of air stones water circulation was formulated for filter feeders after introducing the plankton and unfiltered fresh seawater.

Installation of Three Different Base Covering Tanks

Gravel Base Aquarium

For experimental gravel base aquarium, two filter plates were placed on a glass tank bottom. Filter plate used as an under gravel filter system. For base covering, sufficient quantity of clean and safe gravels spreaded over the tank floor and filter plate, to a good depth, and sloped so that the back is higher than front (a ratio of 1:5 from front to back will allow good filtration and give the aquarium a sense of perspective). In addition to clean and safe decorative stones, rocks, different size of gastropod shells and a arch shaped tile were placed in different places of the tank, which was used as decorative and hiding place for invertebrates (adopted from Mills, 1987).

Sand Base Aquarium

For experimental sand base aquarium, two filter plates were placed on a glass tank bottom. Above the filter plate a nylon mesh was placed for keeping a space between tank floor and top filter plates so as to make sand able to enter into the filter plate gaps. For base covering, sufficient quantity of sieved and washed sand spreaded over the tank floor and filter plate and sloped so that the back is higher than front (a ratio of 1:5 from front to back will allow good filtration and give the aquarium a sense of perspective). In addition to clean and safe decorative stones, rocks, different size of gastropod shells and a arch shaped tile were placed in different places of the tank, which was used as decorative and hiding place for invertebrates (adopted from Mills, 1987).

Uca Crab Aquarium

Uca crabs are semi-terrestrial air breathers and therefore no oxygenation of tank water is required. However, it needs a mudflat in the tank. This tank was prepared to have a deep water section at one end and a shallow water section (mud flat area) at the other end of the tank. For experimental *Uca* crab aquarium, single filter plate was placed on a glass tank bottom (deep water area). Above the filter plate placed a nylon mesh for keeping a space between tank floor and top of the filter plate. Because sand able to enter in to the filter plate spaces. For mud flat covering sufficient quantity of sieved and washed sand was spreaded over the tank floor and filter plate to a good depth. Partially water was changed every day. In this tank salt marsh plant *Sesuvium portulacastrum* L. was planted in mudflat areas (Shallow end area) (adopted from Monks, 2005).

Marine Ornamental Fishes for Compatibility Test

For the purpose of compatibility test, certain numbers of live marine ornamental fishes were collected from the Vellar estuary, and they were transported to the laboratory in an aerated condition. The following marine ornamental fishes are maintained in the tank: *Platax teira* (Bat fish), *Scatophagus argus* (Scat fish), *Terapon jarbua* (Tiger fish), *Monodactylus argenteus* (Mono), *Etroplus suratensis* (Pearl spot), *Hyporhamphus limbatus* (Halfbeaks), *Siganus javus* (Rabbit fish), *Chetonodon patoca* (Milk-spotted pufferfish) and *Heniochus acuminatus* (Butterfly fish/Coachman). The study animals number and size (cm) were point out in (Table 13.1).

Table 13.1: Number of Animals and Size of the Experimental Animals

Sl.No.	*Species Name*	*Number of Animals for Study*	*Size Range (cm)*
1.	*Chrysaora quinquecirrha*	10	6.5–7.5
2.	*Rhizostoma* sp.	10	7–9
3.	*Anthopleura* sp.	10	2.5–3.5
4.	*Lepas anserifera*	10	1–3
5.	*Panulirus homarus*	5	21–24
6.	*P. versicolor*	2	8–9
7.	*Thenus orientalis*	10	6.5–8
8.	*Dromia dehanni*	10	6–7
9.	*Dorippe dorcipes*	10	2–3
10.	*Clappa lophos*	10	5.5–6.5
11.	*Matuta lunaris*	10	2.5–4
12.	*Philyra globosa*	10	2.5–3
13.	*Doclea ovis*	10	2.5–3.5
14.	*Charybdis feriata*	10	8–10
15.	*Uca annulipes*	10	1.5–1.8
16.	*Uca triangularis*	10	1.5–1.8
17.	*Clibanarius clibanarius*	10	6–8
18.	*C. longitarsus*	10	4–5
19.	*Coenobita cavipes*	10	5.5–6.5
20.	*Harpiosquilla melanoura*	10	13–15
21.	*Bursetella leachii*	10	6.5–7
22.	*Perna viridis*	10	9–11
23.	*Salmasis virgulata*	10	5.5–7
24.	*Temnopleurus toreumaticus*	10	2.5–3.5
25.	*Astropecten indicus*	10	5–8
26.	*Stellaster incei*	10	6.5–8
27.	*Luidia maculate*	2	14–17
28.	*Ophiocnemus marmorata*	10	4–15

Laboratory Acclimatization

The experimental animals were acclimatized for laboratory condition, according to Hargreaves (2003) using the drop-by-drop method. The collected specimens are allowed to float with the container in the unlit aquarium for 15 minutes. After that a small quantity of water from the aquarium was added to the container. This process was repeated for four times of 5 minutes interval. For aeration, air stone were used to provide gentle aeration in the container.

Food and Feeding

Commercial pellet feed of ornamental fish and "Wockcee" shrimp feed was used for invertebrates and ornamental fishes. *Artemia* nauplii, was hatched out in laboratory for filter feeders. Live feed such as larvae of freshwater prawn, *Acetes* sp., and fish larvae were collected from Vellar estuary during high tide time with the help of scoop net. Fish and Shrimps were collected from the Parangipettai landing centre, which was used after cleaning. Mussels were collected from the Vellar estuary, which was used after cleaning. Unfiltered fresh seawater collected from the Vellar estuary during high tide time for filter feeders, which contained suspended particles. Planktons were collected from Vellar estuary for filter feeders.

Method for *Artemia* Nauplii Hatching

Artemia cyst (San Francisco Bay Strain) was hatched in a glass jars containing filtered seawater of 35 per cent. 1 g of cyst was added in 1 litre of seawater and was provided with a light source and vigorous aeration. After 18 hours *Artemia* cyst hatched out into nauplii. The freshly hatched nauplii was washed in fresh seawater and offered to the filter feeders like *Lepas anserifera* and Green-Lipped Mussels *Perna viridis.*

Results

Behaviour of Some Ornamental Invertebrates in Captive

Aquarium behaviour studies have been carried out in laboratory aquarium, the observation reveals that the *Chrysaora quinquecirrha* and *Rhizostoma* sp., are delicate for handling, because they have long tentacles and soft bell, it is easily damaged during handling. They float and swim weakly around the aquarium with much of effort by means of muscular contractions of the bell. We can see the engulfed food through the transparent body of the jelly.

If we touch the anemone *Anthopleura* sp., in tank, they contract and withdraw their tentacles. If they are disturbed out of the tank, they will squirt the water out. The *Lepas anserifera* is generally peaceful in tank and the cirri are constantly active. *Panulirus homarus* and *P. versicolor* are slow mover and swift swimmer when it's disturbed. The sand lobster *Thenus orientalis* is sluggish, when disturbed which shows jet like swift backward movement.

Dromia dehanni buries beneath the sand or gravel, *Dorippe dorcipes, Clappa lophos* and *Philyra globosa* buries beneath the sandy bottom only, by a series of backward movement and holding a shell of a large bivalve on dorsal side from aquarium, excluding *Clappa lophos* and *Philyra globosa. Matuta lunaris* some time swims in the aquarium and buries below the sand or gravel by a series of backward scuttles and predatory in aquarium. *Doclea ovis* is active animal and live on substratum. *Charybdis feriata* is an aggressive crab, which attacks other crabs and other invertebrates in tank and should not advisable to maintain it with other crabs.

Uca annulipes and *U. triangularis* does not adapt to permanently being kept underwater because *Uca* crabs are semi-terrestrial air breathers and very few males are injured during contests.

Clibanarius clibanarius and *C. longitarsus* are active mover, which wander the whole tank, aggressive during feeding time and as scavenger. *Coenobita cavipes* is a slow mover and scavengers.

Harpiosquilla melanoura is generally peaceful, which rest on the floor of the tank if disturbed it swims fast. They suddenly take the food when introduced to the aquarium. *Bursetella leachii* is harm less animal and constantly wanders the floor and glass wall of the aquarium and don't discharge the ink in tank. If *Perna viridis* feel touch, which close the shells, further it becomes permanently attached to the aquarium substrata with its byssus threads never move.

Salmasis virgulata is wander in to the whole aquarium but slowly move on the substratum, which attach the gravels and broken shells on the test from the aquarium. *Temnopleurus toreumaticus* is a very slow mover, which is not a very active species compare with *Salmasis virgulata*.

Astropecten indicus, *Stellaster incei* and *Luidia maculate* are peaceful animal and slowly move on the tank floor and which buries under the sand, but slightly dorsal side is visible, excluding *Luidia maculate*. *Ophiocnemus marmorata* is a peaceful animal and buries beneath the sand by a series of arm movement.

Feeding Behaviour and Food Preference

Feeding behaviour and food preferences studies of ornamental invertebrates have been carried out in laboratory aquarium, the observations are reveals that the Jellyfishes *Chrysaora quinquecirrha* and *Rhizostoma* sp., both are feed *Artemia* sp., *Acetes* sp., and minced and squashed fish and shrimp meat in aquarium. *Artemia* and Acetous are mixed with the aquarium water, which is trapped by jellyfish oral a..n. In aquarium it can accept pellet feed also. Oral arm of jellyfish having mucus, which trap the food particles, then moves up to the mouth. Minced and squashed fish, shrimp meat and pellet feeds are attached with oral arm with the help of forceps.

The sea anemone *Anthopleura* sp., captures fish larvae, small freshwater prawn and *Acetes* sp., when close to the tentacles and minced and squashed fish and shrimp meal through hand feeding. In aquarium it can accept pellet feed also and fed once a day is enough. Pellet feeds are placed on the mouth with light touch with the help of forceps then they close the tentacles and engulf the pellet food.

For feeding *Lepas anserifera* and *Perna viridis*, 3 litres of fresh seawater is added to the tank, once a day so as to feed planktons and suspended particles are available in it. In tank 250 ml beaker full newly hatched *Artemia* are poured with vigorous aeration is applied for water circulation. *L. anserifera* also accept very minute, minced and squashed fish meals when placed on the active cirri through forceps except that *Perna viridis*.

In aquarium *Panulirus homarus*, *P. versicolor*, *Thenus orientalis*, *Dromia dehanni*, *Dorippe dorcipes*, *Matuta lunaris*, *Philyra* globosa, *Doclea ovis*, *Uca annulipes*, *U. triangularis* and *Harpiosquilla melanoura* are consumes minced, cleaned and washed fish, shrimp and clam meat through hand feeding, they receives the food through chelipid.

*Clappa lophos*h consumes minced, cleaned and washed fish, shrimp, clam meat and shrimp pellet feed through hand feeding which receives the food through chelipid. *Charybdis feriata* is a voracious feeder and active predator prefers to hunt feed like shrimps and fishes that lie in wait for their prey. In aquarium which consumes minced, cleaned and washed fish, shrimp and clam meat through hand feeding which receives the food through chelipid.

In aquarium *Clibanarius clibanarius*, *C. longitarsus* and *Coenobita cavipes* are accept any food offered and consumes minced, cleaned and washed fish, shrimp and clam meat through hand feeding this receives the food through chelipid and which consumed the missed food particles of others, as a scavenger. *Bursetella leachii* consumes the minute particles on substratum as grazers.

Salmasis virgulata and *Temnopleurus toreumaticus* can also take minced and squashed fish, shrimp meal and accept pellet feed also which are placed on the oral side or any tube feet through pointed forceps. Tube feet get any food particles, which move the food particles towards the oral side with the help of many tube feet.

Long time starved sea stars *Astropecten* indicus, Stellaster *incei* and *Luidia maculate* are breaks the live small mussels and clams are for feeding. It can also take minced and squashed fish, shrimp meal and accept pellet feed also which are placed on the oral side through pointed forceps. The brittle star *Ophiocnemus marmorata* feed on detritus and dead organic materials in aquarium and large specimens accept pellet feed also.

Tank suitability experiment and behaviors was presented in (Table 13.2). Longevity and survival studies of ornamental invertebrates have been carried out in laboratory aquarium, most of the study animals showed the maximum longevity period of about more than hundred days and *Chrysaora quinquecirrha and Rhizostoma* sp. shows the minimum longevity period of about one week (but not exceed two weeks). Some of animals like *Bursetella leachii* and *Ophiocnemus marmorata* shows the longevity period of about up to one month (between 25 to 32 days). *Dromia dehanni* shows the maximum survival rate of about 90 per cent. *Chrysaora quinquecirrha* and *Rhizostoma* sp. shows the minimum survival rate of 0 per cent. The results of longevity showed in (Table 13.3) and rate of survival showed in (Figure 13.1-13.7). Compatibility of marine ornamental invertebrates with fishes and invertebrates like *Anthopleura* sp., *P. globosa, C. lophos, D. dorcipes, L. anserifera, T. orientalis, P. homarus, P. versicolor C. clibanarius, C. longitarsus, Coenobita cavipes, P. viridis, T. toreumaticus, S. virgulata, S. incei, A. indicus, L. maculata* and *O. marmorata* are compatible with both fishes and invertebrates. *C. feriata* not compatible with both fishes and invertebrates. Compatibility studies of ornamental invertebrates have been carried out in laboratory aquarium and the observations were presented in (Table 13.4).

Table 13.2: Suitability for Sand, Gravel Base and *Uca* Crab Aquarium

Sl.No.	Species	Habitats in Wild	Suitable for Sand Base Aquarium	Suitable for Gravel Base Aquarium	Suitable for Uca Crab Aquarium
1.	*Chrysaora quinquecirrha*	Free-floating marine form (Waikiki aquarium, 2000)	✓	✓	–
2.	*Rhizostoma* sp.	Free-floating marine form (Waikiki aquarium, 2000)	✓	✓	–
3.	*Anthopleura* sp.	Attached to the seafloor by an adhesive basal disc (Waikiki aquarium, 1998)	✓	✓	–
4.	*Lepas anserifera*	Marine	✓	✓	–
5.	*Panulirus homarus*	Rocky shallow waters (Dhandapani, 1993)	–	✓	–
6.	*P. versicolor*	Found among the coral reefs and in the sea grassbeds (Dhandapani, 1993)	–	✓	–
7.	*Thenus orientalis*	Sandy bottom (Kizhakudan, 2004)	✓	–	–

Contd...

Table 13.2–Contd...

Sl.No.	Species	Habitats in Wild	Suitable for Sand Base Aquarium	Suitable for Gravel Base Aquarium	Suitable for Uca Crab Aquarium
8.	*Dromia dehanni*	Offshore mud bottom (Sethuramalingam and Ajmal Khan, 1991)	✓	–	–
9.	*Dorippe dorcipes*	Muddy bottom of sea (Sethuramalingam and Ajmal Khan, 1991)	✓	–	–
10.	*Clappa lophos*	Offshore of muddy sand bottom (Sethuramalingam and Ajmal Khan, 1991)	✓	–	–
11.	*Matuta lunaris*	Estuarine and marine shallow sandy beach of intertidal region (Sethuramalingam and Ajmal Khan, 1991)	✓	–	–
12.	*Philyra globosa*	Sandy bottom of sub tidal region (Sethuramalingam and Ajmal Khan, 1991)	✓	–	–
13.	*Doclea ovis*	Deep sandy and muddy bottom of sea (Sethuramalingam and Ajmal Khan, 1991)	✓	–	–
14.	*Charybdis feriata*	Marine, rocks, stones, sandy muddy substratum (Sethuramalingam and Ajmal Khan, 1991).	✓	✓	–
15.	*Uca annulipes*	Burrows in firm mud and sandy mud of estuarine shore (Sethuramalingam and Ajmal Khan, 1991).	–	–	✓
16.	*Uca triangularis*	Burrows in firm mud and sandy mud of estuarine shore (Sethuramalingam and Ajmal Khan, 1991)	–	–	✓
17.	*Clibanarius clibanarius*	Inshore water (Ajmal Khan, 1992)	✓	✓	–
18.	*C. longitarsus*	Estuarine (Ajmal Khan, 1992)	✓	✓	–
19.	*Coenobita cavipes*	Estuarine shore (Ajmal Khan, 1992)	–	–	✓
20.	*Harpiosquilla melanoura*	Marine (Ajmal Khan and Lyla, 2004)	✓	✓	–
21.	*Bursetella leachii*	–	✓	–	–
22.	*Perna viridis*	Estuarine (Toonen, 2003)	✓	✓	–
23.	*Salmasis virgulata*	Soft bottom of sea	✓	✓	–
24.	*Temnopleurus toreumaticus*	Soft bottom of sea	✓	✓	–
25.	*Astropecten indicus*	Sandy mud bottom	✓	–	–
26.	*Stellaster incei*	Sandy shallow bottom	✓	✓	–
27.	*Luidia maculate*	Marine	✓	✓	–
28.	*Ophiocnemus marmorata*	Sandy bottom and associated with medusae	✓	–	–

✓: Suitable; –: Not suitable.

Table 13.3: Number of Animals and Longevity

Sl.No.	Species Name	Number of Animals for Study	Longevity (Days)
1.	*Chrysaora quinquecirrha*	10	Above one week
2.	*Rhizostoma* sp.	10	Above one week
3.	*Anthopleura* sp.	10	Above 100 days
4.	*Lepas anserifera*	10	Up to one month
5.	*Panulirus homarus*	5	Above 100 days
6.	*P. versicolor*	2	Above 100 days
7.	*Thenus orientalis*	10	Above 100 days
8.	*Dromia dehanni*	10	Above 100 days
9.	*Dorippe dorcipes*	10	Above 100 days
10.	*Clappa lophos*	10	Above 100 days
11.	*Matuta lunaris*	10	Above 100 days
12.	*Philyra globosa*	10	Above 100 days
13.	*Doclea ovis*	10	Above 100 days
14.	*Charybdis feriata*	10	Above 100 days
15.	*Uca annulipes*	10	Above 100 days
16.	*Uca triangularis*	10	Above 100 days
17.	*Clibanarius clibanarius*	10	Above 100 days
18.	*C. longitarsus*	10	Above 100 days
19.	*Coenobita cavipes*	10	Above 100 days
20.	*Harpiosquilla melanoura*	10	Above 100 days
21.	*Bursetella leachii*	10	Up to one month
22.	*Perna viridis*	10	Above 100 days
23.	*Salmasis virgulata*	10	Above 100 days
24.	*Temnopleurus toreumaticus*	10	Above 100 days
25.	*Astropecten indicus*	10	Above 100 days
26.	*Stellaster incei*	10	Above 100 days
27.	*Luidia maculate*	2	Above 100 days
28.	*Ophiocnemus marmorata*	10	Up to one month

Discussion

Keeping and maintenance of marine ornamental invertebrates are interesting one than fish keeping. Knowledge about aquarium behaviour, feeding, food preference, longevity, survival and compatibility with fishes and among invertebrates are needed to keep and maintain the ornamental invertebrates in the aquarium. The obtained results are discussed here with the earlier works and observations.

Table 13. 4: Compatible marine ornamental invertebrates and fishes

Sl.No.	Species Name	Compatibility with Fishes	Compatibility with Invertebrates
1.	*Chrysaora quinquecirrha*	–	+
2.	*Rhizostoma* sp.	–	+
3.	*Anthopleura* sp.	+	+
4.	*Charybdis feriata*	–	–
5.	*Matuta lunaris*	+	–
6.	*Doclea ovis*	–	+
7.	*Dromia dehanni*	+	–
8.	*Philyra globosa*	+	+
9.	*Calappa lophos*	+	+
10.	*Dorippe dorcipes*	+	+
11.	*Uca annulipes*	+	–
12.	*Uca triangularis*	+	–
13.	*Lepas anserifera*	+	+
14.	*Harpiosquilla melanoura*	–	+
15.	*Thenus orientalis*	+	+
16.	*Panulirus homarus*	+	+
17.	*P. versicolor*	+	+
18.	*Clibanarius clibanarius*	+	+
19.	*C. longitarsus*	+	+
20.	*Coenobita cavipes*	+	+
21.	*Perna viridis*	+	+
22.	*Bursetella leachii*	+	–
23.	*Temnopleurus toreumaticus*	+	+
24.	*Salmasis virgulata*	+	+
25.	*Stellaster incei*	+	+
26.	*Astropecten indicus*	+	+
27.	*Luidia maculata*	+	+
28.	*Ophiocnemus marmorata*	+	+

The aquarium suitability classified into following ways by Hargreaves (2003):

Aquarium Suitability

- ☆ Easy (good for beginners and experienced aquarists)
- ☆ Moderately easy (good for aquarists with 6 months experience or more)
- ☆ Moderately difficult (only suitable for experienced aquarists)
- ☆ Very difficult (requires optimal aquarium conditions for success)
- ☆ Almost impossible (suitable only for the most experienced aquarist).

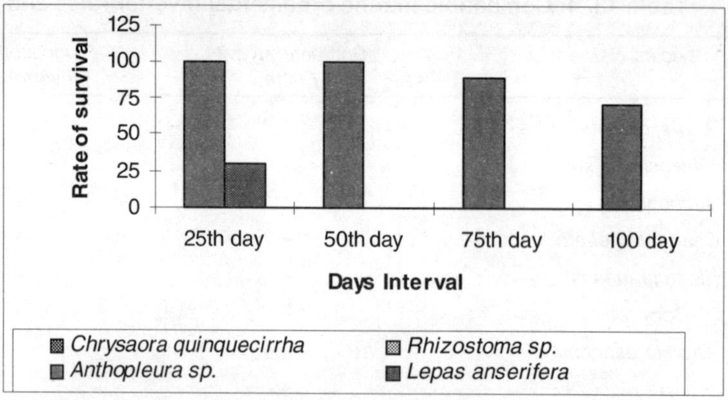

Figure 13.1: The Rate of Survival

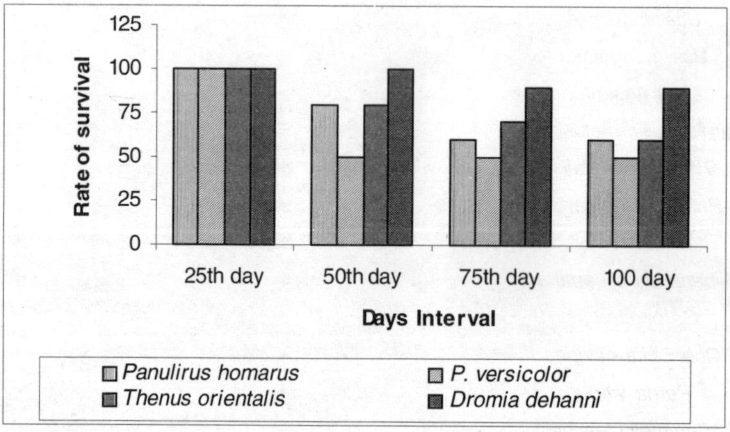

Figure 13.2: The Rate of Survival

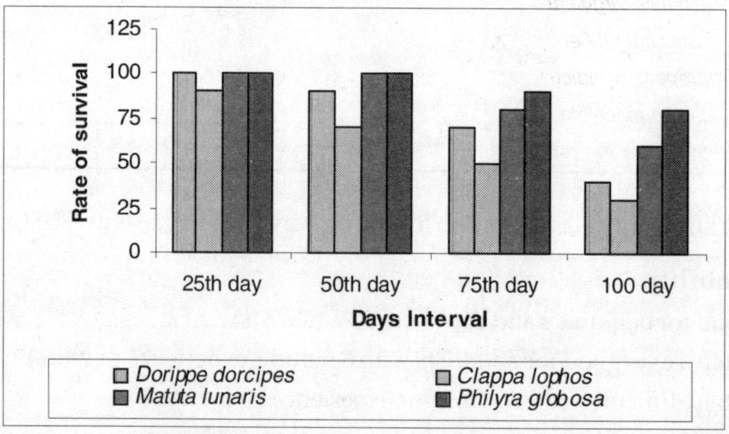

Figure 13.3: Shows the Rate of Survival

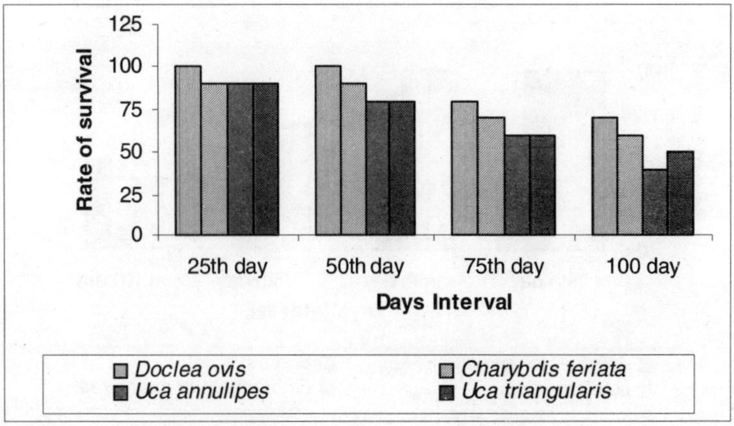

Figure 13.4: Shows the Rate of Survival

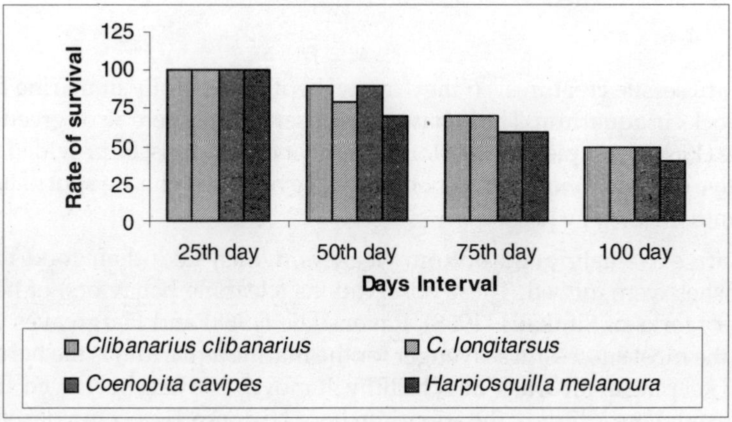

Figure 13.5: Shows the Rate of Survival

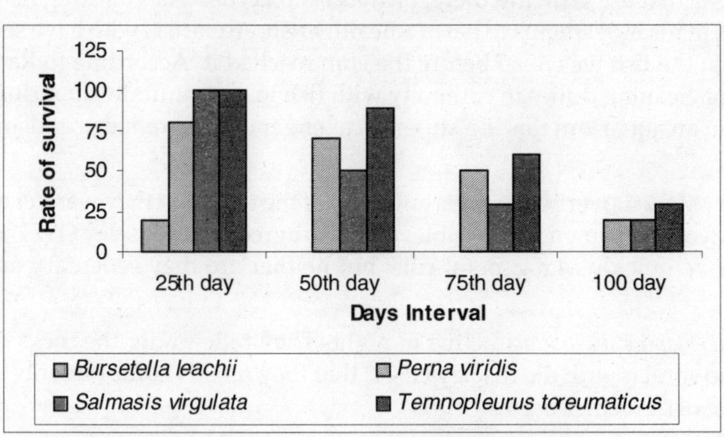

Figure 13.6: Shows the Rate of Survival

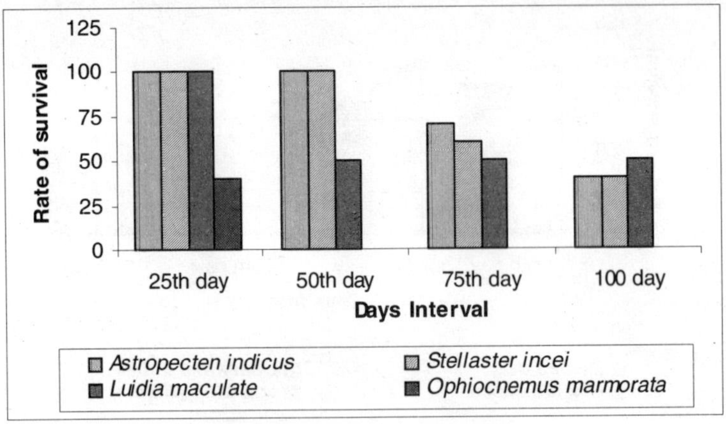

Figure 13.7: Shows the Rate of Survival

Behaviour

Sea anemones are sessile creatures, so they can be kept successfully in marine aquariums with some stones or flat rocks in aquarium. The behavioural observations were also agreed with the earlier works of Ravensdale (1968) that a piece of flat stone or similar objects must be provided in the aquarium for the anemone to live on and, although the anemone will eventually chose a spot of its own, it should be placed directly onto the rock by hand.

Hermit crabs are extremely good bottom cleaners, which take their food that those other invertebrates and fishes were missed. These observations (cleaning behaviors) of hermit crab were similar to the earlier works of Simkatis (1958), Ravensdale (1968) and Hargreaves (2003). Simkatis (1958) denotes that the most successful scavenger for the marine aquarium is the hermit crab. Thus it contributes much in keeping a salt-water tank healthy. It moves carrying borrowed shell from corner to corner of the tank and there is little of the aquarium floor he doesn't cover in a day. Of course, it will eat the food the fishes miss, but make sure he gets his share because he really earns it. It will climb over coral and eat the algae that are growing there. He has been accused of attacking fishes but he is too slow and ponderous in his movements to harm a healthy fish. Aquarists who have seen him carry off a fish can be sure that the fish was dead before the crab reached it. According to Ravensdale (1968), small Hermit crab for cleaning duties in company with fish in a community aquarium. According to Hargreaves (2003), in an aquarium this is a superb scavenger of food remains and an excellent algae controller.

Sea stars are very slow movers on gravels and sand of the tank. But they wander around the tank. During this the tube feets movements are visible. This also agreed with Riseley (1971) observation that starfishes do not move quickly as a general rule, but neither are they sedentary and they wander around the tank.

Brittle star *Ophiocnemus marmorata* active in night. They hide under the rocks during daytime. This observation also similar with the Riseley (1971) that they mostly come out only at night they are by far the most active of the echinoderms.

A much smaller species the purple-tipped sea urchin (*Psammechinus miliaris*) is common on rocky seashores. It is greenish in colour with purple-tipped spines. It has a habit of covering itself with bits of broken shell, seaweed or pebbles, which it holds over itself with its tube feet (Sexton and Wilson, 1969). The similar observation was observed in this present study, sea urchin *Salmasis virgulata* attaches some small gravel, piece of shells, etc. Tube feet of the sea urchins assist in this attachment.

Tank Types

We shouldn't keep the jellyfish sea nettle *Chrysaora quinquecirrha* in stone or rock-decorated aquarium. Because the oral arms of sea nettle is long in length. So, they entangled with stones and rocks.

In sand base aquarium box crab *Matuta lunaris* buries below the sand by a series of backward scuttles. The paddle like leg scoops the sand away underneath and as the crab goes in back wards the sand settles on top of it.

Box crab *Matuta lunaris,* sponge crab *Dromia dehanni,* masking crab *Dorippe dorcipes,* shame face crab *Calappa lophos,* purse crab *Philyra globosa,* sea star *Astropection indicus* and brittle star *Ophiocnemus marmorata* are better kept in plenty of sand base aquarium. *Dromia dehanni, Dorippe dorcipes, Calappa lophos* and *Philyra globosa* buries beneath the sand by a series of backward movement. Sea star *Astropecten indicus* has pointed tube foot, which aids in quick burrowing into sand. The dorsal side is slightly visible above the sand. *Dromia dehanni* also found buried deep in gravel base aquarium. Because they are from sandy and sandy mud bottoms of wild. *Ophiocnemus marmorata* associate with scyphozoan medusae. Even though which is adapt with sand base aquarium without medusae. Observation on *Astropection indicus* was also similar with the James (2003). Thus *Astropection indicus* partially buried them under sand in tank. Slightly the dorsal side is exposed on the sand.

Jackamn (1974) describes the base covering that the most successful covering is one that just obscures the base of the tank, and if the odd crab disturbs this layer and exposes the tank floor it is an easy matter to redistribute it by using a glass rod or small stick. It is better to do this regularly than develop other troubles later through using too thick a layer of sand. Riseley (1971) said generally about, starfishes are found on reefs, reef flats and on muddy or sandy shores, sometimes partly buried in the sand. We would not try to keep them without plenty of coral rock and sand.

In wild sand lobster *Thenus orientalis* substratum is generally sandy, reported by Kizhakudan *et al.,* (2004). So keep this species in sand base aquarium. In gravel base aquarium sea urchin *Salmacis virgulata* attaching gravels and bits of shells. *Ophiocnemus marmorata* buries beneath the sand by a series of arm movement. A few handfuls of beach sand is recommended by Simkatis (1958), because it is beneficial to some marine animals. Sand should be boiled and dried before being used in the aquarium.

Short stalked goose barnacle *Lepas anserifera* as sedentary crustaceans that spend life, fixed on logs, timber, plastics, mettle cans, bulbs, buoys, bottles, etc. So we should keep with the substratum in aquarium. If the substratum floats, fix them between the decorative stones.

During collection the sea anemones were found fixed with the *Doclea ovis* and hermit crabs. Even in tanks we should keep them with host animals. If you want to keep the anemones solitary, remove the anemones without any damage from the host. Then place the anemones on the stones. They gradually firmly attach with the stones or rock. In aquarium arch like tile provides cave like structure for crabs. This observation was same with the earlier works of Simkatis (1958) and Jackman (1974). According to Simkatis (1958) anemones should be removed carefully from the lump of coral or other surface to which they are attached. They can be handled with the fingers but usually a sharp knife is needed to remove them without damage. If the anchorage is a small stone or a bit of coral, take it along with the specimen. Jackman (1974) describes, the anemones can be quite at home on rock ledges, and a small cave can be provided for the shore crabs. A few ledges or overhanging rocks will provide a stamping ground for the mussels.

Monks (2005) explains the *Uca* crab aquarium: *Uca* spp are small crabs in various shades of yellow and brown, but the males have one enormously developed claw with which they display. These are semi-terrestrial animals and will not adapt to permanently being kept underwater. They need rocks, a sandbank, or some other sort of structure above the water level onto which they can climb and feed. Ideally, keep in a mudskipper style aquarium.

Land hermit crab *Coenobita cavipes* have adapted to life on the shore and intertidal areas. So *Uca* crab aquarium is ideal for this land hermit crab.

Feeding Behaviour and Food Preference

Several diets (Pellet feed and live feed) have been tested in different form of invertebrate species, the feeding behaviour and the suitability of the food preferences to be species dependent.

In aquarium jellyfishes feed on plankton, *Artemia* sp., Acetous and minced and squashed fish and shrimp meat. In aquarium it can accept pellet feed also. According to Jackman (1974), feeding jellyfish is no easy matter unless plentiful supplies of plankton are available. Sometimes a specimen will readily take minute pieces of crushed mussel that are stirred into water, but this feeding method brings water pollution.

The piece of food should be placed on the oral disc of anemone, with the help of forceps, no need to push the food into the oral disc. The feeding in sea anemones can be initiated by disturbing the tentacles with a small forceps. This feeding behaviour observation was similar with the previous works of Hargreaves (2003) that a piece of food should be placed on the oral disc, midway between the margin of the disc and the mouth. Never try to force feed an anemone by pushing food into the oral opening itself. This will block the siphonoglyphs necessary for hydrostatic control and water/oxygen exchange within the creature's body.

After food is placed, the tentacles cover the food and pass it to oral disc or mouth. The unwanted food and digested particles are removed through mouth of the anemone. This observation was agreed with the earlier observations of Ravensdale (1968), in his view the tentacles immediately begin to close and an almost magnetic force holds the intended food in place. The tentacles then pass the food along the line until it is forced into the mouth. Unappreciated food will be regurgitated along with the natural wastes in the form of a slime coated ball, which is ejected at intervals from the mouth.

Sea anemones can live for weeks without food, weekly twice feeding is enough, which take pellet feed also in aquarium. This feeding observation was agreed with the earlier works of Ravensdale (1968) and Mills (1987). According to Ravensdale (1968), it can live for weeks without any food and rarely needs feeding more than twice weekly. Mills (1987) denotes, invertebrates, such as sea anemones, need not be fed as frequently as fish; they may take in food when it is offered, but sometimes they spit out any surplus food or digest and then eject the entire contents of the stomach, almost as soon as your back is turned. This can also pollute the aquarium. Depending on their size, sea anemones have different feeding requirements. The smaller species need small particles of food-proprietary planktonic foods, cultured rotifers, brine shrimp, small freshwater daphnia, cyclops and maybe tubifex can all be used, and also proprietary aquarium granular and tablet foods. Feeding twice a week is usually sufficient for large species, but smaller species can be fed with, say, brine shrimp a little more often. Needless to say, sea anemones should be offered food only when their tentacles are fully expanded. Be sure to remove any partially digested foods, ejected after a few hours or the next day, to prevent tank pollution.

In captive sea anemones feed with live *Acetes* sp., small shrimps, which are introduced in large number. Some times live feeds entangled in the sea anemones tentacles. So they don't enough for the good survival of the sea anemone. So we should give supplement feed (small and squashed fish and shrimp meat) also for sea anemone as Simkatis (1958) statement in his view, sea anemones are animals and have to be fed. In their natural habitat, they stretch their graceful tentacles and catch their food in the currents of incoming and outgoing tides. They are thus able to keep well and healthy. In the aquarium, however, where food is not constantly available, the anemone should be hand fed. Tiny bits of shrimp well rinsed under a running faucet) or beef should be dropped in the center of the flowerlike creature.

Simkatis (1958), reported about hermit crabs food seeking: If you look at him very carefully, you will notice it has two bright blue eyes mounted on stalks. This does not locate the food; it depends on its sense of smell for this function and he described the hermit crabs as a "Pet". He will go hungry if he doesn't have you looking out for his interests. When he learns you are his friend, he will take food from your fingers. If a scrap of food is left unmolested on the aquarium floor, he will eventually find it and at times he literally falls over backward in his enthusiasm to stake his claim. This similar activity was observed in the present study also.

In aquarium filter feeders like short stalked goose barnacle *Lepas anserifera* and Green-Lipped Mussels *Perna viridis* consume newly hatched nauplii of *Artemia* sp. After introducing the live feeds, vigorous aeration is need for water circulation. In order to visualize the feeding apparatus we should place the animals with substratum very close to the front glass and a magnifying glass is fixed. This observation was similar with the earlier work of Mills (1987) that the newly hatched nauplii stages are ideal for filter feeding invertebrates.

Place the small bit of food on ventral side of the sea star with the help forceps, which take. Throughout this study sea stars don't break any bivalve shells for feeding. This observation also agreed with the earlier work of Ravensdale (1968) that the method of feeding adopted by the starfish is as remarkable as most of its other characteristics. Its powerful arms are quite capable of forcing open the largest of shells and, as explained previously, instead of putting food into the stomach the starfish puts its stomach into the food. Riseley (1971) said about sea star, in the wild these animals feed by using the tube feet to force open bivalves and then inserting the stomach.

Brittle stars are scavengers, they consume the uneaten small particles remains in the tank floor. This feeding behaviour agreed with Riseley (1971) general statement on brittle stars that they are such wonderful scavengers and also it is fascinating to see the thin arms waving at feeding time. The aim seems to be to capture pieces of chopped shrimp without exposing their bodies. They can seize food with the arms and then pass it along the underside to the mouth using the tube feet as a conveyor belt.

Longevity and Survival

Lifespan of jellyfish is usually only about one week (7-9) in captive, which are gradually shrinks in size. Jellyfishes, *Lepas* and ragged sea hare are lead up to one month in tank. Jellyfishes are difficult to keep in captivity for any length of time and require special care. *Lepas* are filter feeder, so enormous amount of plankton are need to thrive. Feeding of ragged sea hare is difficult one. For filter feeders get more amounts of plankton and suspended food they not thrive more than one month. So further detailed study is needed to keep this in aquarium.

Jackman (1974) statement regarding to, *Aplysia punctata* can be kept with varying degrees of success, but are definitely not hardy in small tanks with limited circulation of fresh sea water. Feeding of ragged sea hare *Bursetella leachii* is a difficult one in captive. They need algal growth in tank. With out algal growth they meet starvation. But the filter feeder green mussel *Perna viridis* live more than 100 days in tank as a hardy species. Brittle star *Ophiocnemus marmorata* is a hardy species but is a delicate for handling. So they easily go away the aquarium.

Compatibility of Marine Ornamental Invertebrates with Fishes and among Invertebrates

Jelly fishes *Chrysaora quinquecirrha* and *Rhizostoma sp.* are floating and wanderers of aquarium. The tentacles of *Chrysaora quinquecirrha* have long tentacles. So, the fishes and crabs disturb them. We should keep the jellyfishes in the separate tanks. The angel crab *Charybdis feriata* is an active predator. It hunts the other invertebrates and fishes in the tank. So, we should keep this separate tank. Large size of Box crab *Matuta lunaris* holds the small size of the box crab *which* them chelate legs. We should keep separate the large and small size *Matuta lunaris* in tank. Spider crab *Doclea ovis* having elongated second pair of leg. In tank during movement it is attract the fishes. They try to attack them. Male *Uca* crabs contest with other males in some time so they lose their large chelipede. Ornamental mudskippers are suitable for *Uca* crab aquarium. This is agreed with the Monks (2005). Mantis shrimp *Harpiosquilla melanoura* attacks the fishes so we should keep this in a separate tank. Number of hermit crab is considered as suitable for aquarium. According to Simkatis (1958), do not take too many hermit crabs but plan to have two for each aquarium with a capacity over forty gallons and one each for those having less. We have taken hermits that were about the size of a pea but perhaps that size of marbles would be more effective. They grow fairly rapidly but it takes well over a year before they become objectionably large. By the time they reach ping-pong-ball dimensions, they have become pets and their release back into the sea has its ceremonial aspects.

Ragged sea hare *Bursetella leachii* is a soft animal. Crabs attack this in tank. So, we should keep this in separate tank. Tiger fish *Therapon jarbua* is a predator. So, it is not compatible with invertebrates and fishes.

These findings will surely enable to keep and maintain the ornamental invertebrates in aquarium.

Acknowledgements

The authors are greatful to Prof. Dr. T. Balasubramanian, Director, CAS in Marine Biology, Annamalai University, Parangipettai for providing the facilities to carry out the work and encouragement.

References

Ajmal Khan, S. and Lyla, P.S., 2004. Stomatopod: The new potential resource. *Ocean-Life, Food and Medicine, Expo*, pp. 1–12.

Ajmal Khan, S., 1992. *Hermit Crabs of Parangipettai Coast*. CAS Marine Biology Publication, Annamalai University, pp. 39.

Dhandapani, P., 1993. *Handbook on Aqua Farming: Identification of Cultivable Species*. MPEDA Publication, Kochi, pp. 45–49.

Hargreaves,V., 2003. *The Complete Book of the Marine Aquarium*. Greenwich Editions, London, pp. 294.

Jackman, L.A.J., 1974. *Seawater Aquaria*. Douglas David and Charles Ltd, Canada, pp. 176.

James, D.B., 2003. *Marine Faunal Diversity of India: Training Manual.* Marine Biological Station, Zoological Survey of India, Chennai, pp. 332–340.

Kizhakudan, J.K., Thirumilu, P. and Manibal, C., 2004. Fishery of the sand lobster *Thenus orientalis* (Lund) by bottomset gillnets along Tamil Nadu coast. *Mar. Fish. Infor. Serv.*, T and E., 181: 6 and 7.

Mills, D., 1987. *The Practical Encyclopedia of the Marine Aquarium.* Salamander Books Ltd, London, UK, pp. 208.

Monks, N., 2005. *Brackishwater Aquarium,* [Online]. Available: http://www.aquariacentral.com/faqs/brackish/FAQ1.shtml.

Ravensdale, T., 1968. *Coral Fishes: Their Care and Maintenance,* 2 nd Revised Edn. Hope Printers Ltd., London, pp. 255.

Riseley, R.A., 1971. *Tropical Marine Aquaria: The Natural System.* George Allen and Unwin Ltd., Fakenham, London, pp. 186.

Sethuramalingam, S. and Ajmal Khan, S., 1991. *Brachyuran Crabs of Parangipettai Coast.* CAS in Marine Biology Publication, Annamalai University, p. 93.

Sexton, E.W. and Wilson, D.P., 1969. *Illustrated Guide: The Plymouth Aquarium.* Latimer Trend and Co. Ltd., Plymouth, pp. 36.

Simkatis, H., 1958. *Salt-water Fishes for the Home Aquarium.* I.J.B. Lippincott Company, New York, pp. 254.

Strickland, J.D.H. and Parsons, T.R., 1972. *A Practical Handbook of Seawater Analysis.* Bull. Fish. Res. Bd., Canada, 167: 310.

Susheelan, C., Neelakanta Pillai, N., Radhakrishnan, E.V., Rajan, K.N., Manmadhan Nair, K.R., Sampson Manickam, P.E. and Saleela, K.N., 1996. *Handbook on Aqua Farming: Shrimp, Lobster and Mud Crab.* MPEDA Publication, Kochi, pp. 47–54.

Toonen, R., 2003. *Green-Lipped Mussels.* [Online]. Available: http://www.advancedaquarist.com/issues/may2003/invert.htm.

Waikiki Aquarium–Education Department: Marine Life Profile: *Moon Jelly*, 2000. [Online].Available: http://www.waquarium.mic.hawaii.edu./MLP/root/ pdf/marinelife/Invertebrates/Cniderians/Moonjelly.pdf.

Waikiki Aquarium–Education Department: Marine Life Profile: *Sea Anemones*, 1998. [Online].Available: http://www.waquarium.mic.hawaii.edu./MLP/root/pdf/marinel ife/Invertebrates/Cniderians/SeaAnemones.pdf.

Chapter 14

A Note on Water Quality in Aquarium

☆ *Indranil Ghosh*

"Aquariculture" is the term, which exclusively means the cultivation of ornamental fishes within the aquarium system. The aquarium is a miniature aquatic ecosystem where all the necessary components are maintained artificially adopting proper biological and engineering technology. The fishes breed there to produce fry and are reared up to the marketable size with the aid of proper feeding and micro-ecosystem management.

Like the table fishes, the "living jewels" in the aquarium, need utmost care in regard to water quality. Clean water is a must for living fishes, but so-called too much pure water is unsuitable because either they do not have the essential minerals required for fish, or, they may not have buffering capacity to stabilize pH. Distillations, Reverse osmosis, Deionization are the three basic process to produce pure water.

An aquarium usually need clean and clear water, devoid of any suspended particulate matter. Rain water is theoretically pure, but is suitable for a number of high value fishes like Angel, Goldfish, Tetra, Gourami, barbs etc. after proper conditioning.

Whatever the source of water may be, it should be properly filtered and checked for its parameters required for individual groups of fishes. pH and hardness, often with alkalinity, may be regarded as the blood-life for maintaining the *"Glass-Box-Lives"* (Glass-Box=Aquarium).

Natural and spontaneous breeding is the common practice to breed ornamental fish within captive condition. Though some fishes like Sharks along with a few Cyprinids, Loaches, some Murrells, eels, *Ompak* sp. etc. require induction by using spawning hormone. Heteroplastic pituitary extract and other new generation spawning agents like Ovaprim, Ovatide, Pimozize etc. have also been found to be effective. But, proper water chemistry should be maintained for their successful breeding, hatching and, of course, rearing. Except these few, other fishes must require microenvironment manipulation. The Characid fishes like Tetras are highly sensitive to water pH, some of them require acidic medium. We have also found that lowering of pH is an inducing factor for the spawning of high-priced fish like

Neon Tetra, Cardinal Tetra etc. The "Aquarium-Queen" Angel fishes are also fond of a little acidic pH for breeding and rearing. Goldfish and barb also do well in slightly acidic pH. Most of the catfish, cichlids, fighting fish prefer slightly alkaline pH. pH can also control the percentage of male populations in some live bearing fishes like swordtail and platy, though they breed well in alkaline condition. As the male swordtail bearing sword-like extension of their caudal fin, bring much high price in the market than the female fish, the aquarist should maintain proper pH whenever required. pH is also responsible for proper coloration in fish. Hardness is another important parameter for breeding almost all ornamental fish. Hardness also affects fish coloration. Proper hardness triggers the maturity of live bearing ornamental fishes in aquarium. Sometimes, egg-laying brood-fish also need hardness alteration for their final sexual maturity. Temperature is one of the vital factors throughout the life-cycle of any fish. Gestation period of live-bearing fish varies inversely with the ambient temperature. It is well known to us that metabolic activities are mostly dependant on temperature.

Proper nitrogen cycle is a must for a well balanced aquarium, which ensures sufficient amount of beneficial bacteria, responsible for breaking down of complex organic compound. Nitrate, and ammonia are very much toxic to the fishes, and the invertebrates can't tolerate even a little concentration of ammonia in marine aquarium, as evidenced from our daily experience. The aquarists can safely use some products, available in market that can kick the ammonia out of water leaving healthy atmosphere for fish and other invertebrates.

Plants also helps in maintaining proper quality of water, if judiciously used. They not only use CO_2 and provide O_2 but also utilize fish excreta as fertilizer. Some plants can absorb heavy metals and other toxicants from water injurious for fish.

The scavenger organisms like snails are of immense benefit for the aquarium. They eat the surplus feed along with the faecal matters and detritus. They also graze upon the red and green algal matters formed upon the surface of glass and other objects kept within aquarium. A bivalve, if properly used, can also engulf the particulate matters.

Proper illumination is necessary for smooth running of photosynthesis and also ensures the required photoperiod for the organisms.

Chlorine and sometimes chloramines are sometimes added in tap-water to make it suitable for human use, but it may be highly toxic for aquatic lives. So dechlorination is needed before using the water in aquarium, otherwise, it may be the 'silent-killer' of fish.

Chapter 15

Water Borne Diseases

☆ *Deepa Dev*

Introduction

Medical science has evolved through many stages in last few centuries. From the 'Germ theory of diseases' propounded in 1860 by Louis Pasteur, the real reason of infectious diseases was identified.

The last century can be labeled as a century of infectious diseases. However, it is interesting to note that the control of infectious diseases has started long before antibacterial were identified.

The two classic examples which showed that control of infectious diseases for community was nothing but environmental sanitation control are the famous cases of Typhoid Mary and of Panama canal construction. The incidences showed the role of environmental engineering in the control of diseases.

This generated into a totally new thinking in medicine with a paradigm shift from cure to prevention of diseases.

Of the five natural resources Jal (water), Aap(agni) Pruthvi (land), Vayu (wind) and Tej (light) are capable of creating illness and catastrophe if the attention and respect is not given to them.

In the Indian context water is labeled as Jeevan *i.e.*, life. It is purest of pure washing all sins and protecting us.

The personal hygiene has long been emphasized in India. The practices of daily baths sometimes more than once, clean washed clothes, amblution, all show the emphasis. However, the true concept of environmental sanitation is nowhere seen.

With the advent of urbanization and many people using same resources of water, this lack of environmental sanitation has been catastrophic. Just as water is Jeevan it can become equivalent to death it is true in periods of floods and famine but it is still true in periods of calm just because we do not take care of it. Water borne diseases constitute one important group of illnesses which can easily be prevented however

- 200 million people in India do not have access to safe water.
- 1.5 million Children less than 5 years die annually due to water borne diseases.
- 200 million person days of work are lost.

This statistic talks of water borne diseases however condition is till grim if we consider following definitions in the context of water and diseases.

Definitions

Water Borne Diseases

Disease transmission occurs by drinking contaminated water transmitted by faeco oral route *e.g.*, Typhoid, hepatitis and cholera etc.

Water Related or Water Breeding Diseases

Refers to infections spread by insects that depend on water. Insect vectors breeding in water transmit malaria, filaria, dengue, yellow fever etc. The infection may also occur by inhalation to microbes on water droplets such as those produced by showers, air conditioning, irrigation etc.

Water Washed or Wasted Diseases

Includes infections on the outer body surface due to lack of sufficient quantity of water for washing and personal hygiene *e.g.*, Trachoma, skin ulcers scabies typhus etc.

Water Based Diseases (Non Faeco Contamination)

Refers to the infections transmitted through aquatic invertebrate animals *e.g.*, *Schistosomiasis, Dracunculiasis.*

Classification

Caused by Presence of infective agent

- *Viral*: Hepatitis A and E, Poliomyelitis, rotavirus diarrhoea,.
- *Bacterial*: Typhoid, paratyphoid, bacillary dysentery, *E. Coli* diarrhoea, Cholera.
- *Protozoal*: Amoebiais, giardiasis.
- *Helminthic*: Round worm, thread worm and hydatid disease.
- *Leptospiral*: Weils disease.

Caused by Presence of Aquatic Host

- Snail: Schistosomiasis,
- *Cyclops*: Dracunculiasis. Fish tape worm.

Dracunculiasis

Asia is free since Feb 2006. In India last case was detected in July 1996. Feb 2001 India is described free from guinea worm disease.

Disease Burden of Water Borne Diseases

Disease	*World*	*India*
Hepatitis A and E	10–15 persons/lakh/yr 10–25 per cent cases in children 1–5 per cent cases in adults	Exact incidence not known
Poliomyelitis	1593 lab confirmed cases	114 lab confirmed cases
Diarrhoea	3.2 episodes/child/yr Mortality in <5–4.9/1000 3 million deaths	Contribute to 1/3rd of global burden
Cholera	Cases–131000 Deaths–2272 CFR–1.72	Cases–3156 Deaths–6 CFR–0.19
Typhoid	17 million affected > 6 lakh deaths/yr	Cases–653580 Deaths–1417
Amoebiasis	Prevalence 2–60 per cent 70000 deaths/yr	Affects 15 per cent of the Indian population
Ascariasis	1.3 billion affected	Prevalence is 250 million
Hookworm	Cases–151 million Deaths–65000	>200 million affected
Leptospirosis	High prevalence in tropical countries	Since 1999 cases found in many states

Table 15.1: Water Borne Viral Diseases

Epidemiology	*Hepatitis A and E*	*Poliomyelitis*	*Viral Diarrhoea*
Agent	Enterovirus type 72 Picornaviridae family	RNA Virus 1,2,3	Rota, adeno, astro,calci, corona, norvak, entervirus
Reservoir	Human cases	Human cases	Animals, humans
Infective material	Faeces	Faeces, oropharyngeal secretions	Faeces
Communicability	2 weeks before and 1 week after appearance of jaundice	7-10 days before and after symptoms	–
Age	All common in children	6 months to 3 years	6 months–3 years
Sex	Equal	3:1 M:F	Equal
Risk factors	–	Fatigue, trauma surgery, injections	Malnutrition, prematurity, incorrect feeding
Environmental factors	Heavy rainfall, poor sanitation, overcrowding, contaminated food and water, flies July–September		
Incubation period	15–45 days	7–14 days	
Clinical spectrum	Fever, chills, headache fatigue followed by anorexia, nausea vomiting dark urine and jaundice	In apparent 91–96 per cent Abortive 4–8 per cent Non paralytic 1 per cent Paralytic <1 per cent	Loose motions, vomiting, fever and dehydration
Prevention	Control of reservoir Control of transmission Control of susceptible population by immunoglobulin and vaccines	Immunization IPV and OPV	Appropriate clinical management, better MCH practices sanitation, health education, immunization and fly control

Table15. 2: Water Borne Bacterial Diseases

Epidemiology	*Cholera*	*Typhoid*	*Diarrhoeal Diseases*
Agent	V. Cholerae or, classical and El tor Toxin producing	S. Typhi and Paratyphi A and B	*E. Coli, Salmonella shigela B. cereus, Copylobacter Jejuni*
Reservoir	Cases and carriers	Cases and carriers	Animals, humans
Infective material	Stools, vomit of cases and carriers	Faeces, Urine water food, finger flies	Faeces
Communicability	Temporary 7–10 days Chronic 10 years	Temporary and chronic	Variable
Age	All	5-19 years	6 months-3 years
Sex	Both	Males	Equal
Risk factors	Population mobility		Malnutrition, prematurity, incorrect feeding
Environmental factors	Human habits favoring water and soil pollution, low standard of personal hygiene lack of education, poor quality of life.		
Incubation period	Few hours to five days	10-14 days	Different depends on causative agent
Clinical spectrum	Three stages, Evacuation, collapse and recovery	Chills fever malaiseeadache cough sore throat pea soup diarrhoea, splenomegaly	Loose motions, vomiting, fever and dehydration
Prevention	Sanitation, chemoprophylaxis, vaccination and health education	Control of cases and carriers immunization	Appropriate clinical management, better MCH practices sanitation, health education, immunization and fly control

Water Borne Diseases: Prevention and Control

I) Controlling the Reservoir

1. Early Diagnosis

The first step in control of diseases is rapid identification. Frequently laboratory procedures may be required to confirm the diagnosis.

2. Notification

Infectious diseases are notifiable by law under international health regulations. This is to prevent an outburst of epidemic by putting in place control measures and can be compared to stamping out the spark rather than calling the fire brigade to put out the fire.

3. Epidemiological Investigation

Investigation of epidemic is an important tool in general health. This identifies and helps in taking proper care the steps of epidemiological investigation include

(*i*) Verification of diagnosis

(*ii*) Confirmation of existence of epidemic

(*iii*) Defining population at risk

(*iv*) Rapid search for all cases and characteristics

Table 15.3: Water Borne Parasitic and Other Diseases

Epidemiology.	*Amoebiasis*	*Ascariasis*	*Hookworm*	*Leptospirosis*
Agent	Entamoeba histolytica	Ascaris Lumbricoides	A. duodenale N Americans	Spirochetes 23 serotypes
Reservoir	Man	Man	Man	Goat, Sheep Cattle mice etc.
Infective material	Faeces	Faeces containing fertilized eggs	Faeces containing ova of hook worm	Urine of infected animals
Communicability	Several years	Until all fertile females are destroyed	As long as person harbours parasite	
Age	Any	Children	15-25	Children
Sex	Both	Male	Both	Male
Environmental factors	Low socioeconomic status, poor sanitation, use of night soil for agriculture purposes		Sandy damp soil More than 40 inches rainfall	Poor housing limited water supply inadequate methods of waste disposal
Incubation period	2-4 weeks	2 months	7 and 5 weeks	10 days
Clinical spectrum	Mild abdominal discomfort, diarrhoea, dysentery; extra intestinal	Nausea, abdominal pain cough, passage of live worm in stools and vomitus	Anaemia, growth retardation LBW babies abortion still birth	Mild fever of severe and fatal disease with kidney and liver involvement
Prevention	Primary prevention: Sanitation food hygiene health education Secondary prevention Early diagnosis and treatment			

(*v*) Data analysis

(*vi*) Formulation of hypothesis

(*vii*) Testing of hypothesis

(*viii*) Evaluation of ecological factors

(*ix*) Further investigation of population at risk

(*x*) Writing a report

4. Isolation

Isolation is defined as separation for the period of communicability of infected persons or animals from others in such places and under such conditions as to prevent the direct or indirect transmission of infectious agent from those infected to those who are susceptible.

The period of isolation for different diseases depends on period communicability.

5. Treatment

Though treatment in strict sense will benefit the patient, in cases of infectious diseases it has an added advantage that it reduces the communicability, duration of illness and secondary cases.

6. Quarantine

Is defined as limitation of freedom of movement of such well persons and domestic animals exposed to communicable disease for a period of time not longer than the longest incubation period.

(II) Interruption of Transmission–Sanitary Barrier

It mainly concerns prevention of Faeco oral route of transmission.

The faecal-oral route of disease transmission

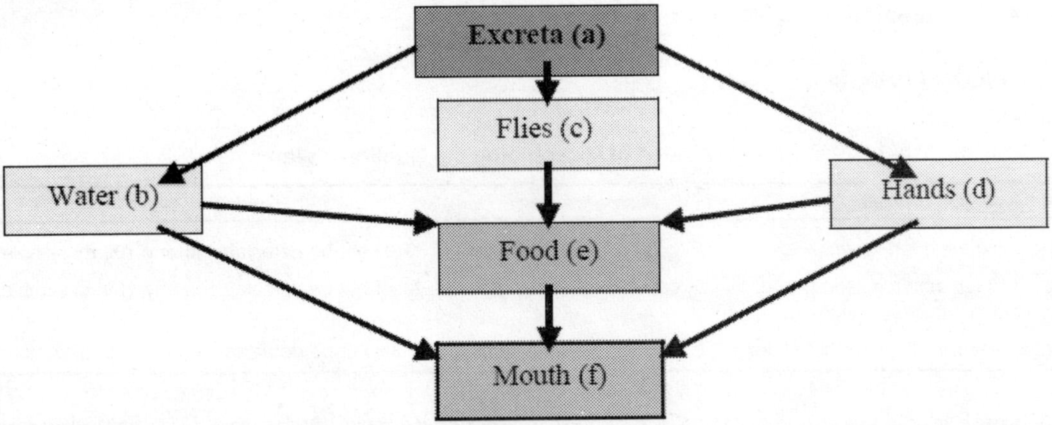

(a) To (b): Improper excreta disposal, leads to contamination of natural water resources in river, streams, ponds, lakes, and aquifers. Also contamination of water supply line with sewage system contaminates water.

(a) To (c): Improper excreta disposal leads flies and other insects to contact with faecal matters. Surface of insects harbours pathogenic microorganisms. Poor solid waste management enhances breeding of flies.

(a) To (d): hands get contact with microorganisms and remain on the surface due to improper hygiene practice or due to poor water supply.

(b) To (e): Untreated contaminated water transmits microorganisms to food, while washing raw vegetables, cooking and mixing. Irrigation with contaminated water makes vegetables and fruits potential source of microorganisms. Some microorganisms are resistant to high temperature (eggs of some parasites and intestinal worms), so uncooked or improperly cooked food contaminated with microorganisms may remain source of infection.

(c) To (e): When flies and other insects contaminated with micro organisms sit on food or get cooked with food become source of infection.

(d) To (e): unclean hands transmit microorganisms to food while handling.

(b) To (f): If untreated contaminated water is consumed.

(e) To (f): If raw or improperly cooked food contaminated with microorganisms are consumed.

(d) To (f): If food is eaten by using unclean hands

For prevention of water borne diseases this chain of faecooral transmission should be broken (Sanitation barrier). For this purpose Water should be *easily accessible, adequate in quantity, free from contamination safe and readily available through out the year.*

(III) The Susceptible Host

A. Provision of Safe and Wholesome Water

It is the basic step in prevention of waterborne diseases. Safe and wholesome water is defined as as free from pathogenic agents, free from harmful chemical substances, pleasant to taste *i.e.* free from colour and odour and usable for domestic purposes. It should fulfill to following criteria's.

The guidelines for drinking water quality recommended by WHO relate tofollowing variables

1. Acceptability aspects
2. Microbiological aspects
3. Chemical aspects
4. Radiological aspects

Microbiological Aspects

Bacteriological

Guidelines Value of Microbiological Quality of Water

Type of Water	*Microorganism*	*Guideline Value*
All water intended for drinking	*E. coli* and other coliform	Must not be detectable in any 100 ml sample
Treated water entering the distribution system	*E. coli* and other coliform	Must not be detectable in any 100 ml sample
Treated water in distribution system	*E. coli* and other coliform	Must not be detectable in any 100 ml sample

Though faecal streptococci and *Cl. perfringens* are also bacterial indicators *E. coli* are chosen as indicators of faecal pollution because of

1. Ease of detection
2. Abundance in human intestine
3. Loner survival
4. Greater resistance to natural purification.

Virological Aspects

Drinking water should be free from any viruses infectious for man.

Parasiotolgical Aspects

Drinking water should not contain any protozoa or helminthes or eggs. *Chemical, radiological and acceptability aspects are not discussed.

B. Immunization

People at risk are protected by one or more of the following strategies

1. Active Immunization

Active immunity is resistance developed by an individual as a result of antigenic stimulus. When such an antigenic stimulus is given deliberately and artificially it is called active immunization. Vaccination is common means of active immunization amongst waterborne diseases vaccines against Poliomyelitis, hepatitis, cholera, typhoid are available. The problem with active immunization against water borne disease is

(*a*) Immunity is short lived.

(*b*) The strains of many viral pathogens undergo mutation.

2. *Chemoprophylaxis*

(*a*) It is defined as protection from or prevention of disease using medicines. It is available against cholera, typhoid, etc. It is given in cases where person is going to be exposed to disease risk for short duration of time. Like travel in endemic areas.

(IV) Prevention at Community Level

1. Water Supply

Water should be *easily accessible, adequate in quantity, free from contamination safe and readily available through out the year.*

Easily Accessible

For drinking purpose water should be available within 1.6km in plain area and 100 meters in hilly area.

Adequate in Quantity

In urban areas	140-150 ltrs/capita per day (LPCD)
In rural areas	40 LPCD
Drinking	03
Cooking	05
Bathing	15
Washing utensils	07
Ablution	10
Animal	30 LPCD

One hand pump for 250 person rate 12 ltrs/min

Free from Contamination

Discussed above.

For this purpose various methods of purification are in use to make water from these sources safe and potable. These are:

Large scale

- Storage
- Filtration: Rapid or mechanical filters
  Slow sand or biological filters
- Disinfection Chlorination
- Ozonation
- UV irradiation

Small Scale

- Boiling
- Chemical disinfection
- Filtration

2. Sanitation, and Health Education

1. Protecting food from flies interrupts the faeces-flies-food route (at a household level).
2. Increasing the quantity of water available.
3. Changing hygiene behaviour
4. Hand washing
5. Care in disposing of faeces.
6. Proper use of latrines by adults and children.
7. Proper use and maintenance of water supply and sanitation systems.
8. Proper maintenance of pumps and wells. Pipes and taps should always be kept clean.
9. Disposing of organic and animal wastes properly to avoid flies.

National Health Programmes

1954:	National water supply and sanitation programme
1972-73	Accelerated rural water supply (ARWSP)
1980-1990	International drinking water supply and sanitation decade
1986	National drinking water mission (technical mission renamed in 1991 as Rajiv Gandhi National drinking water mission).
1986	National water policy
1993	Accelerated urban water supply
2003	Swajaldhara
2005	Creation of drinking water infrastructure as a part of Bharat Nirman

Important Days in calendar for health awareness regarding water

22nd March	World water day
17-23 August	World water week
15th Oct	Global hand washing day

Inequalities in water availability is a measure of in equal development. Access to safe and adequate drinking water is a fundamental right.

Investigation of an Epidemic

1. Verification of diagnosis
2. Confirmation of existence of epidemic
3. Defining population at risk
4. Rapid search for all cases and characteristics
5. Data analysis
6. Formulation of hypothesis
7. Testing of hypothesis
8. Evaluation of ecological factors
9. Further investigation of population at risk
10. Writing a report

Chapter 16

Studies on Aquatic Insects in Relation to Physico-chemical Parameters of Anjani Reservoir in Sangli District of Maharashtra

☆ *S.A. Khabade and M.B. Mule*

ABSTRACT

Aquatic insects and some important abiotic factors of Anjani water reservoir and their inter-relationships were studied in 2004 to 2005. Both insect abundance and abiotic parameters showed fluctuations from month to month. The monthly study of water quality and insect fauna of Anjani water reservoir has been carried out to investigate the impact of various meterological, physical and chemical parameters on distribution and abundance of the insects.

In the present investigation, the water quality parameters such as light transparency, total solids, total dissolved solids, total suspended solids, water pH, electrical conductivity, total alkalinity, acidity, hardness, magnesium, calcium, chloride, residual chlorine, dissolved oxygen, free CO_2, hydrogen, sulphide, sodium, potassium, nitrate and phosphate may not cause any adverse effect on the distribution and abundance of insect fauna, but rainfall, air temperature, humidity and water temperature were the most effective parameters which vastly affected the insect population of the pond.

Introduction

Water quality plays an important role in the growth of aquatic animals and their distribution and abundance. The water quality standards below and above the optimum level may lead to either stress or death among the aquatic animals. The water quality mainly depends on physical, chemical and microbiological parameters. Sometimes, water quality may be changed due to organic and chemical pollution of water body.

Aquatic insects comprise an ecologically important group of organisms in freshwater systems. They are known to play a very important role in the processing and cycling of nutrients, as they belong to several specialist feeding groups such as filter feeders, deposit collectors, scrapers, shredders as well as predators (Lamberti and Moore, 1984). Their importance as biomonitor or indicators of freshwater pollution has also been amply demonstrated (Wiederholm, 1984; Metcalfe, 1989).

Aquatic insects spend part of all their lives in soil or water exhibit special structural and behavioural adaptations to the physico-chemical and biotic conditions found in each. These insects form an important link in the nutritional cycle of an aquatic ecosystem, as they constitute a bulk of food for fishes. Extensive work has been done on the seasonal hydrobiological conditions of freshwater bodies of India (Sinha and Sinha, 1999; Kaushik *et al.*, 1990; Pandey, *et al.*, 1992; Singh, 1993; Arvind Kumar, 1994).

Most of the waterbodies have been unmindfully used for the disposal of waters as dust-bin far beyond their assimilative capacities and have been grossly polluted. The domestic sewage and industrial effluents contain pollutants that cause harmful effect on receiving waters and adverse impact on human health as well as aquatic biota (Telliard and Robin 1987; Kumar 1996b).

Since aquatic insects are bound to an aquatic habitat for most part of their life-cycles, any change in their number and composition in the population at a given time and space may indicate a change in the water quality. Benthic insects are considered as promising organism for use in diversity biomonitoring because of their case of collection, large number of species and sensitivity to water quality (Roy and Sharma, 1983; Kumar 1995a). As such many of aquatic insects form biological indicators of the environmental quality. Thus, by studying them it is possible to anticipate the impact of pollution even before drastic physical and chemical changes have occurred.

Information on the inter-relationship between aquatic insect population and physico-chemical conditions is scanty. Hence, the influence of some physico-chemical factors on the abundance of aquatic insects of Anjani reservoir of Tasgaon tahsil of Sangli district has been studied.

The environmental pollution affects the general quality of our surroundings and poses risk to our health and wellbeing. The contaminants present in the earth in one way or other get collected through the streams and rivers and ultimately reach the ocean. Making assessment of water quality and devising control strategies are not static, but an ongoing process. A number of factors act concurrently in a lake and in turn on the quality of the water (Churchill and Buckingham 1956). Deshmukh (1964) has studied the physico-chemical characteristics of Ambazari lake in Nagpur, Maharashtra. The water quality and conservation aspects of five water bodies in and around Hyderabad, Andhra Pradesh, are discussed by Kodarkar (1995). Similarly, a number of studies on physico-chemical and biological quality of the waters have been extensively carried out (Busulu *et al.*, 1967; Chakraborty *et al.*, 1977; Adwant, 1989; Khatavkar and Trivedy, 1992; Joshi and Bisht, 1993 and Gill *et al.*, 1993). Considerable work has been carried out on water quality assessment of freshwater bodies by using the species diversity in other regions of India (Sharma and Rai 1991; Karim 1993).

Material and Methods

The aquatic insects were collected monthly during July 2004 to June 2005 period from the Anjani water reservoir with an insect collecting net made up of nylon cloth having mesh size: 40-80 cm^2. The samples were cleaned and preserved in 5 per cent formalin. Then the insects were identified in the laboratory with the help of standard literature of Tonapi (1959); Michael (1973), Macan (1959).

Monthly collection of water sample was done from the period July 2004 to June 2005, by using samplers *i.e.* plastic containers of 5 litre size. The sites selected for water sampling are site I (SI) and site II (S II) which lies near the dams earthen embankment and the feeders canal which recharge the reservoirs water respectively. The meterological parameters such as air temperature and humidity were determined in the field. Similarly, few physico-chemical parameters like water temperature, pH and dissolved oxygen were determined in the field during monthly visits. The electrical conductivity, total alkalinity, hardness, magnesium, calcium, chlorides, residual chlorine, acidity, free CO_2, hydrogen sulphide, sodium, potassium, nitrate and phosphate were analysed in laboratory, using standard methods of water analysis described by Trivedy and Goel (1984) and APHA, AWWA, WPCF (1985).

Results and Discussion

The results on water quality assessment in terms of a number of physico-chemical parameters are summarized in the Table 16.3 and 16.4.

In the present investigation it has been found that the species diversity was higher during monsoon when rainfall was maximum. Table 16.1 shows the rainfall data during July 2004 to June 2005. During rainy season water temperature was also optimum which affect the species diversity and it becomes higher. During summer, in the month April 2005 when water temperature was high about 31°C, the insect species diversity was minimum in this month. Rao (1976) also opined that a heavy rainfall in monsoon period increases the diversity while the poor rainfall has adverse effect on the diversity index of hemipteran insects. Julka (1977) has also suggested that in the complexes of interdependent factors governing the seasonal variations in diversity of aquatic bugs, temperature and rainfall appear to be the important factors. In the present investigation it is also observed that the species diversity was minimum during summer. In the situation, only pollution tolerant aquatic insects will be present and pollution intolerant species decline. The pollution tolerant species can grow more rapidly without competition for space, nutrients, predation and other extrinsic and intrinsic factors too. This results in heavy dominance of these species leading to the decline in the values of species diversity and also in the evenness of species (Cairns, 1977).

Table 16.1: Month Wise Record of Rainfall in mm During July 2004 to June 2005 at Reservoir Site

Sl.No.	*Month*	*Year*	*Rainfall in mm*	*Name of the Block*	*Water Reservoirs Coming Under Savlaj Block*
1.	July	2004	59.6	Savlaj Block	Siddhewadi
2.	Aug.	2004	75.00		Anjani
3.	Sept.	2004	94.20		Balgawade
4.	Oct.	2004	14.00		Bastawade
5.	Nov.	2004	–		
6.	Dec.	2004	–		
7.	Jan.	2005	–		
8.	Feb.	2005	–		
9.	Mar.	2005	–		
10.	April	2005	–		
11.	May	2005	–		
12.	June	2005	14.00		

–: No Rainfall.

Table 16. 2: Solubility of Oxygen in Pure Water Exposed to Water-saturated Air at Mean Sea Level Pressure of 760mm Hg

Temperature (°C)	*DO (mg/liter)*	*Temperature (°C)*	*DO (mg/liter)*
0	14.16	18	9.18
1	13.77	19	9.01
2	13.40	20	8.84
3	13.05	21	8.68
4	12.70	22	8.53
5	12.37	23	8.38
6	12.06	24	8.25
7	11.76	25	8.11
8	11.47	26	7.99
9	11.19	27	7.86
10	10.92	28	7.75
11	10.67	29	7.64
12	10.43	30	7.53
13	10.20	31	7.42
14	9.98	32	7.32
15	9.76	33	7.22
16	9.56	34	7.13
17	9.37	35	7.04

Source: Handbook of Common Methods in Limnology (Lind O. T. 1974).

In present study, the rainfall was recorded during July 2004 to October 2004 and also during June 2005. The higher rainfall about 94.2 mm was reported during September 2004. In the remaining months no rainfall was recorded in the Savlaj Block. The air temperature and humidity reported during rainy season was optimum. The maximum air temperature of about 38°C was reported during month May 2005 and minimum air temperature of about 26°C was reported during months December 2004 and January 2005. The maximum humidity 79.00 per cent was reported during February 2005 while minimum humidity 31.00 per cent was reported during January 2005. The optimum range of air temperature and humidity was also responsible for maximum insect species diversity during rainy season.

Light transparency of water depends on the total solids, total dissolved solids and total suspended solids of the water. In present study it is found that these physical parameters not cause any adverse effect on the distribution and abundance of insect diversity, in Anjani water reservoir during study.

In present investigation the maximum water temperature about 31°C was recorded during April 2005 and minimum water temperature about 24°C was recorded during December 2004 and January 2005. The maximum light transparency about 136.0 cm was recorded during October 2004 while minimum light transparency about 16.0 cm was recorded during July 2004. The maximum total solids reported was 5215 mg/L, during May 2005 while minimum total solids reported was 700 mg/L, during October 2004. The maximum total dissolved solids reported was 3200 mg/L during June 2005

Table 16.3: Meterological Parameters of Anjani Water Reservoir from July 2004 to June 2005

Sl.No.	Month and Year → / Parameters ↓ Sites →	July 2004 S_1	July 2004 S_2	August 2004 S_1	August 2004 S_2	Sept. 2004 S_1	Sept. 2004 S_2	Oct. 2004 S_1	Oct. 2004 S_2	Nov. 2004 S_1	Nov. 2004 S_2	Dec. 2004 S_1	Dec. 2004 S_2	Jan. 2005 S_1	Jan. 2005 S_2	Feb. 2005 S_1	Feb. 2005 S_2	Mar 2005 S_1	Mar 2005 S_2	April 2005 S_1	April 2005 S_2	May 2005 S_1	May 2005 S_2	June 2005 S_1	June 2005 S_2
1.	Air Temp. (°C)	31	31	32	33	33	34	31	32	29	28	27	26	26	26	30	30	32	32	37	36	35	38	32	31
2.	Humidity %	72	67	72	65	73	62	52	52	52	45	52	52	50	31	73	79	73	46	36	41	67	65	55	55

All values are mean of four readings.

Table 16.4: Meterological Parameters of Anjani Water Reservoir from July 2004 to June 2005

Sl.No.	Month and Year → / Parameters ↓ Sites →	July 2004 S_1	July 2004 S_2	August 2004 S_1	August 2004 S_2	Sept. 2004 S_1	Sept. 2004 S_2	Oct. 2004 S_1	Oct. 2004 S_2	Nov. 2004 S_1	Nov. 2004 S_2	Dec. 2004 S_1	Dec. 2004 S_2	Jan. 2005 S_1	Jan. 2005 S_2	Feb. 2005 S_1	Feb. 2005 S_2	Mar 2005 S_1	Mar 2005 S_2	April 2005 S_1	April 2005 S_2	May 2005 S_1	May 2005 S_2	June 2005 S_1	June 2005 S_2
1.	Air Temp. (°C)	31± 0.88	31± 0.88	32	33	33	34	31	32	29	28	27	26	26	26	30	30	32	32	37	36	35	38	32	31
2.	Humidity %	72	67	72	65	73	62	52	52	52	45	52	52	50	31	73	79	73	46	36	41	67	65	55	55

All values are mean of four readings.

while minimum total dissolved solids reported was 336 mg/L during November 2004. The maximum total suspended solids reported was 3210 mg/L during July 2004 while minimum total suspended solids reported was 100 mg/L during January 2005. The analysis of historical records reveals a number of inter-relationship among key water quality parameters used in the assessment of cumulative impacts. The recommended concentration of TSS is 80 mg/L.

In present study it is also found that the chemical parameters such as pH of water, electrical conductivity, total alkalinity, hardness, magnesium, calcium, chloride, dissolved oxygen, acidity, residual chlorine, free CO_2, hydrogen sulphide, sodium, potassium, nitrate, phosphate may not cause any adverse effect on the distribution and abundance of insect fauna.

References

Adwant, M.P., 1989. Limnological studies on Godavari basin at Nanded, Maharashtra, India. *Ph.D. Thesis,* Marathwada University, Aurangabad.

APHA, AWWA, WPCF, 1985. *Standard Method for the Examination of Water and Wastewater*, 16th Edn. American Public Health Association Inc., New York, p. 1268.

Busulu, K.R., Arora, H.C. and Aboo, K.M., 1967. Certain observations on self purification of Kher river and its effect on Krishna river. *Ind. J. Environ. Hlth,* 9(4): 275–296.

Cairns, J.J., 1977. Indicator species by the concept of community structure as amindes of pollution. *Water Resources Bulletin,* 10: 338–347.

Chakraborty, R.D., Roy, P. and Singh, S.S., 1977. A quantitative study on plankton and physico-chemical conditions of river Jamuna at Allahabad in 1945. *Ind. J. Fish,* 6(1): 186–203.

Churchill, M.A. and Buckingham, R.A., 1956. Statistical method for analysis of stream purification capacity. *Sewage Ind. Wastes,* 28: 517–537.

Deshmukh, S.B., 1964. Physico-chemical characteristics of Ambazari lake water, Nagpur, Maharashtra. *Ind. J. Environ. Hlth.,* 6(3): 166–188.

Gill, S.K., Sahota, G.P.S. and Sahota, H.S., 1993. Phytoplankton and physico-chemical parameters: Examination of river Sutlej. *Ind. J. Environ. Pro.,* 13(3): 171–175.

Joshi, B.D. and Bisht, R.E., 1993. Some aspects of physico-chemical characteristics of western Ganga Canal near Jwalapur, Haridwar. *Himalayan J. Env. Zool.,* 7(1): 76–82.

Julka, J.M., 1977. On possible fluctuation in the population of aquatic bugs in a fish pond. *Oriental Insects,* 11: 139–149.

Karim, S.W., 1993. Use of species diversity in the evaluation of water quality of two ponds. *Columban J. Life Sci.,* 1: 27–29.

Kaushik, S., Sharma, S.M.N. Saxena and Saksena, D.N., 1990.Abundance of insects in relation to physico-chemical characteristics of pond water of Gwalior (M.P.). *Proc. National Acad. Sci., India,* 60(B)II.

Khatavkar, S.D. and Trivedy, R.K., 1992. Water quality parameters of river Panch Ganga near Kolhapur and Ichalkaranji, Maharashtra, India. *J. Toxicol. Env. Monit.,* 2(2): 113–118.

Kodarkar, M.S., 1995. Conservation of lakes IAAB Publ. No. 2 IAAB, Hyderabad, pp. 82.

Kumar, A., 1995a. Population dynamics and species diversity of odonate larvae in the wetlands of Santhal Pargana, India. *Proc. Nat. Sci.,* 65: 265–278.

Kumar,A., 1996b. Impact of Industrial pollution on the population status of *Plantanista gangetica* in the river Ganga in Bihar, India. *Pol. Arch. Hydrobiol.*, 43: 469–476.

Kumar, Arvind, 1994. Role of species diversity of aquatic insects in the assessment of population in wetlands of Santhal Parganas (Bihar). *J. Environment and Pollution*, 1 (3 and 4): 117–120, *J. of Freshwater Biol.*, 8: 241–146.

Lamberti, G.A. and Moore, J.W., 1984. Aquatic insects as primary consumers. In: *The Ecology of Aquatic Insects*, (Eds.) V.H. Resh and D.M. Rosenberg. Praeger Publishers, New York, p. 164–195.

Macan, T.T., 1959. *A Guide to Freshwater Invertebrate Animals*. Longman Group Ltd., London.

Metcalfe, J.L., 1989. Biological water quality assessment of running waters based on macroinvertebrate communities: History and present status in Europe. *Env. Poll.*, 60: 101–139.

Michael, R.G., 1973. *A Guide to the Freshwater Organisms*. J. Madurai Univ. Suppl., 7: 23–36.

Pandey, B.N., Jha, A.K. and Lal, R.N., 1992. Benthic macro-invertebrate communities as indicators of pollution in river Mahanand Katihar. *Oikossy*, 9(1 and 2): 35–29.

Rao, T.K.R., 1976. Bioecological studies on some aquatic Hemiptera : Nepidae. *Entomon.*, 1: 123–132.

Roy, S.P. and Sharma, U.P., 1983. Studies on the role of insects in freshwater ecosystem. In: *Proc. Symp. in. Ecol. and Environ. Manages*, pp. 187–191.

Sharma, U.P. and Rai, D.N., 1991. Seasonal variation and species diversity of Coleopteran insects in fish pond of Bhagalpur.

Singh, U.N., 1993. Studies on food and feeding behaviour of selected aquatic insects in artificial habitat. *J. Comp. Physiol. Ecol.*, 18(2): 69–71.

Sinha, K.K. and Sinha, D.K., 1999. Seasonal variations in biomass and production of aquatic insects in Derlict pond and a managed fish pond of Munger, Bihar. *Env. and Ecol.*, 8(4): 1231–1234.

Telliard, W..^.. and Rubin, M.B., 1987. Control on pollutants in wastewater. *J. Chromatographic Science.* 25: 322–327.

Tonapi, G.T., 1959. Studies on the aquatic insects fauna of Poona (Aquatic–Heteroptera). *Proc. Nat. Insti.Sci. India*, 25(6): 321–332.

Trivedy, R.K., Goel, P.K. and Trisal, C.L., 1987. *Practical Methods in Ecology and Environmental Science.* Enviro Media Publication, Karad.

Weiderholm, T., 1984. Responses of aquatic insects to environmental pollution. In: *The Ecology of Aquatic Insects*, (Eds.) V.H. Resh and D.M. Rosenbery. Praeger Publishers, New York, p. 508–557.

Chapter 17

Studies on Groundwater Quality of Latur City in Maharashtra

☆ *M.V. Lokhande, K.G. Dande, S.V. Karadkhele, D.S. Rathod and V.S. Shembekar*

ABSTRACT

Drinking water quality of ground water of Latur at fourteen different places has been studied. Several parameters including, pH, conductivity, dissolved oxygen, free CO_2, total hardness, total alkalinity, chloride, salinity and total dissolved solids are studied. All the values of samples found little higher than the normal value.

Introduction

The ground water is most preferred water recent days. Once believed to be safe from pollution as it is available many strata below the surface, is now proved to be prone to pollution by many researchers across the world. The contamination of ground water may due to improper disposal of domestic and industrial waste water.

The ground water is used for domestic, industrial and agricultural purposes, which add contaminants to the ground water reservoirs. Now a days increasing effects of pollution and over exploitation of ground water have become a serious threat. Therefore it is essential to study the quality of ground water. Many workers such as Praharaj, *et al.* (2004), Tripathi (2003), Sambasiva Rao (1997), Mariappan *et al.* (2000), Shrivsta (1998), Pulle *et al.* (2005), Suryawanshi *et al.* (2004) have been carried exhaustive study on ground water quality.

The present investigation was undertaken to study of physico-chemical characteristics of different areas of ground water from different areas of Latur city. The ground water of Latur city was collected from different zones for physico-chemical analysis of water according to the population and demand for domestic use of that area. Most of the people fulfill their need from ground water. The characteristics such as pH, conductivity, dissolved oxygen, free CO_2; total hardness, total alkalinity, chloride, salinity and total dissolved solids were studied.

Materials and Methods

Samples were collected in properly cleaned polyethylene bottles. The bottles were rinsed thrice with the sample water and collected slowly till bottles were filled completely. Mouth of each bottle was capped and sealed by adhesive tape. Samples were collected from the 14 different selected sites. Immediately after collection of samples each sample bottle was labeled and brought to the laboratory for analysis. The pH was measured using standard pH meter. The conductivity was measured using standard conductivity meter. The further physico-chemical analysis was carried out according to standard methods suggested by APHA (1985), Trivedy and Goel (1984) and IAAB (1998).

Results and Discussion

The variation in physico-chemical characteristics of ground water are given in Table 17.1

Table 17.1: Physico-chemical Parameters of Different Groundwater Samples of Latur District

Sl.No	*Name of the Sampling Sites*	*pH*	*Conductivity*	*Dissolved Oxygen*	*Total Hardness*	*Total Alkalinity*	*Chloride*	*Salinity*	*TDS*
1.	Pochma Galli	8.0	100.5	10.0	200	250	142	260.5	950
2.	Patel Nagar	8.0	74.87	5.6	160	250	142	260.5	540
3.	Sai Road	7.0	116.30	8.4	120	180	284	521.14	1100
4.	Medical College	7.0	119.15	9.2	120	220	411.8	755.65	170
5.	HUDCO Colony	7.0	148.20	7.6	160	100	71	130.28	1575
6.	MIDC Area	8.0	62.66	11.6	215	250	240	267.6	540
7.	Dnyaneshwar Nagar	8.0	84.69	4.8	300	200	270	495.4	720
8.	Babhalgaon Road	8.0	78.49	7.0	180	250	191.7	351.7	1400
9.	Gajanan Nagar	7.0	98.69	12.4	120	150	113.6	208.4	1070
10.	Akshay Nagar	7.5	91.70	9.0	120	230	291.1	533.9	1400
11.	Narayan Nagar	8.0	97.35	10.6	200	400	383.4	703.5	1600
12.	Khori Galli	8.0	92.15	13.6	110	250	220.1	403.8	1370
13.	Parivar Society	8.0	90.55	10.4	124	240	326.6	599.3	970
14.	Ambajogai Road	7.0	98.40	12.0	240	230	276.9	508.1	1820

All values are expressed in mg/lit expect pH and conductivity.

pH

pH is the measure of the intensity or alkalinity measures the concentration of hydrogen ion in water. It does not measure total acidity or alkalinity. In the present study PH value ranged in between 7.0 to 8.5 (WHO, 1994) and maximum permissible limit is 6.5 to 9.2 (ICMR, 1975). The low pH does not cause any harmful effect (Broominathan and Khan, 1994). The pH values were within the drinking water standard. It shows slightly alkaline trend.

Conductivity

Conductivity is an index to represent the total concentration of soluble salts. The conductivity values were found to be varied from 62.66 to 119.15 um/cm. The conductivity value was below the permissible limits as per WHO (1994). Kemmers (1980) reported that infiltered ground water is chemically close to rainwater and has low electrical conductivity. The normal acceptance range is of water up to 1000 mhos/cm (WHO, 1994). The values of tested water are under acceptable range.

Dissolved Oxygen

The value of different ground water samples were found within 4.8 o 13.6 mg/l. Dissolved oxygen of drinking water adds taste and it is a highly fluctuating factors.The permissible standard of dissolved oxygen is above 5 mg/l (Perk and Park, 1980). Dissolved oxygen was found above the permissible limit.

Hardness

Hardness of water samples varies from 110 to 300 mg/l. The values of total hardness of all samples were within the permissible range. The highest desirable limit of total hardness is 300 mg/l (ICMR, 1975). Hardness is known to make an adverse effects on health. Drinking water showes is moderately hard (50 to 150 mg/l). Hardness above 200 mg/l of water is not suitable for domestic use in washing and cleaning. Soft water with hardness of less than 100 mg/l may have lower buffer capacity and more corrosive for water popes (WHO, 1994). Some evidences has been given to indicate its role in heart diseases (Poter, 1974).

Total Alkalinity

The phenolphthalein alkalinity of the water samples is zero but total alkalinity was found between 100 to 400 mg/l. The ISI range of total alkalinity is between 50 to 200 mg/l. The total alkalinity value of Pochamagilli, Patelnager, Babalgaon road and Khori gilli is little higher than ISI range. It may be due to contamination by leaching process through surface water. Pandey and Sharma (1999) reported that the alkalinity is itself not harmful to human being.

Chloride

The chloride content of different water samples of the study areas ranges from 71 to 411.8 mg/l. The highest desirable limit of chloride is 250 mg/l (ICMR, 1975). For the samples of sai road, Dnyaneshwar nager, Akshay nager and Ambajogai road is little higher than WHO and ICMR. But remaining samples are above the permissible limit. It produces salty taste at 250 to 500 mg/l (Trivedy and Goel).

Salinity

The salinity content of different water samples ranged between 130.28 to 755.65 mg/l. All values of salinity are found the above permissible limit of (WHO, 1994).

Total Dissolved Solids

The total dissolved solids of water samples ranged from 170 to 1820 mg/l. The ISI standard for dissolved solid up to 500 mg/l and the maximum permissible quantity is 1500 mg/l (WHO, 1994). The TDS value of the sample water of the selected places is above permissible limit.

Conclusion

Different parameters of water samples of study area do not exceed standard and maximum permissible limit except alkalinity, chloride and total dissolved solids which is more than the prescribed limit of WHO and ICMR. Alkalinity is not harmful to human being (Pandey and Sharma, 1999). Ground water of Latur city is not highly contaminated and can be used for domestic purpose in relation to above studied parameters. Use of ground water is cost efficient because it saves enormous cost of water treatment of surface water. We could suggest that these samples can be used for drinking before proper treatment.

Acknowledgements

The authors are thankful to Principal Dr. R. L. Kawale and Dr. D. G. Solunke, Head department of Zoology and Fishery Science, Rajarshi Shahu college, Latur for providing necessary facilities.

References

APHA, 1985. *Standard Methods for Examination of Water and Wastewater*, 17th Edn. American Public Health Association, New York.

Boominathan, R. and Khan, S.M., 1994. Effects of distillery effluents on pH, dissolved oxygen and phosphate content in Uyyakundan channel water. *Env. Ecol. and Cons.*, 17(4): 850–853.

Dhemere, A.J., Pondhe, G.M. and Singh, C.R., 1998. Groundwater characteristics and their significance with special reference to public health in Paravara area (M.S.). *Poll. Res.*, 17(1): 87–90.

Indian Council of Medical Research, 1975. Manual of standard specification for drinking water, DOC, CDC, New Delhi.

Kemmers, R.H., 1986. *Calcium as Hydrobiological Characteristics for Ecological States*. Tech. Bull., Institute for Land and Water Management Research (ICW), Wageningen, The Netherlands, 47: 13.

Mariappan, P.V., Yegnaraman, V. and Vasudevan, T., 2000. Groundwater fluctuation withn water table level in Thiruppathur block of Sivagangi district (T.N.) *Poll. Res.*, 19(2): 225–229.

Perk, J.E. and Park, K., 1980. *Textbook of Preventives and Social Medicine*, 8th Edn. Messrs Banarsidas Bhanot, Jabalpur, India.

Praharaj, A.K., 1974. In: *Industrial Pollution*, (Ed.) N. Irvingsax. Van Norstandard Reinhold Company.

Pulle, J.S., Khan, A.M., Ambore, N.E., Kadam, D.D. and Pawar, S.K., 2005. Assessment of ground water quality of Nanded city. *Poll. Res.*, 24(3): 657–660.

Rashid, A.K., 1982. Biological assessment of the pollution of four heaviely polluted rivers based on microinvertebrates. Abst, *Nat. Env. Cons.*, New Delhi, p. 9.

Sambasivarao, T., 1997. Sustainability of ground water for irrigation in chandragiri block Distt. (A.P.). *Ecology*, 11(10): 1–9.

Surywanshi, M. Bhagwan, Kalyankar, K.B. and Pande, B.N., 2004. Groundwater analysis in an industrial zone Chikalthana (Aurangabad). *Poll. Res.*, 23(4): 649–653.

Tripathi, J.K., 2003. Groundwater hydrochemistry in and around Bhanjabjhar, Ganjam District (Orissa). *Poll. Res.*, 22(2): 185–188.

Trivedi, R.K. and Goel, P.K., 1984. *Chemical and Biological Methods for Water Pollution Studies*. Environmental Publication, Karad, India.

WHO, 1984. *International Standards for Drinking Water*, Geneva.

Chapter 18

Studies on Oxygen Levels and Temperature Fluctuation in Dhanegaon Reservoir in Osmanabad District of Maharashtra

☆ *M.V. Lokhande, D.S. Rathod, V.S. Shembekar and K.G. Dande*

ABSTRACT

The temperature is one of the most vital ecological factors which control the physiological behaviour and distribution of living organisms. The water temperature is determining factors in the seasonal distribution of organism. The dissolved oxygen concentration is one of the fundamental and important factors influencing the aquatic environment both chemically and biologically. Dissolved oxygen effects the nutrients availability resulting change in the productivity of the entire water body. Dissolved oxygen is one of the most important factor for the metabolism of all aquatic organisms.

In the present investigation an attempt was made to study the fluctuation of dissolved oxygen levels and temperature in Dhanegaon reservoir located at Dhanegaon, Dist. Osmanabad, Maharashtra. In the present study monthly variations of temperature recorded were 20.3 to 30°C, 21.4 to 30.4°C and 20.8 to 30.3°C at spot A, B and C respectively. The monthly variation of dissolved oxygen levels were found to be 8.2 to 12.6 mg/lit, 7.2 to 12.4 mg/lit and 7.6 to 13.5 mg/lit at spot A, B and C respectively. The variation in temperature and dissolved oxygen level in the present investigation were found to be well within the permissible limits. The study was carried out for a period of one year (June 2003 to May 2004) at three sampling station.

Introduction

Lakes, reservoirs, rivers and streams have been critical to the establishment of civilizations throughout human history. Water bodies are essential to humans not only for drinking purposes but also for transportation, irrigations, energy production, industry and west disposal. About 80 per cent of earth surface is covered by water but the inland freshwater availability is account for less than one percent.

Today many reservoirs and lakes in India contaminated run off water from expanding urban and agricultural areas, air pollutants, and hydrologic modification such as drainage of wet lands are just few of the many factors that continue to degrade surface waters. Water is needed for our life at every activities it has become the first responsibility to maintain the quality of water and to conserve the freshwater aquatic environment.

The present investigation has been undertaken to study the various parameters of Dhanegon reservoir which is within the permissible limit prescribed by WHO and ICMR.

The Dhanegaon reservoir is large sized reservoir constructed on Manjara River near village Dhanegaon, Taluka – Kallamb, Dist–Osmanabad. It is basically constructed for irrigation purpose but in the recent year's rapid population explosion and rapid industrialization so this reservoir water supply to Latur, Kaij, Ambajogai and Kallamb for drinking purposes. The reservoir constructed in 1980 lies between 1823 to 1855 N latitude, 7515 to 7615 E longitude. Dhanegaon reservoir has a catchments area of about 2371 Sq. Km. The 73 villages of Beed, Osmanabad and Latur is suppose to be beneficiaries of this project.

Materials and Methods

Water samples are collected in separate wide mouthed crew capped, air tight, opaque polythene container on monthly basis for a period of one year June 2003 to May 2004 at three sampling stations named as spot A, B and C. The spot A is located one side of reservoir and spot B located at the Dhanegaon camp near about 6 Km. away from spot A the spot C is located in the reservoir.

The temperature of lake water samples was recorded in the field itself with the help of centigrade thermometer °C. The amount of dissolved oxygen in the collected water samples was estimated by wrinkle's titrometric method in the laboratory (APHA, 1995 and IAAB, 1998).

Results and Discussion

In the present investigation the maximum temperature of Dhanegaon reservoir was recorded in May it was 30.0°C, 30.4°C and 30.3°C at spot A, B and C respectively in the year 2003-2004. The fluctuation pattern of water temperature is expressed in Table 18.1 and graphically represented in 18.1. The temperature fluctuate in month to month but the seasonal fluctuation of temperature the minimum temperature was recorded in the rainy season and maximum temperature was recorded in the summer season but moderate recorded in the winter season. The temperatures of any area vary from sunlight to shade and from daylight to dark. In Dhanegaon reservoir the maximum and minimum temperature recorded is depend upon the brightness and shadowness of sunshine duration. The water temperature is determining factor for the distribution of aquatic organisms (Allern, 1920) and the variation in water temperature may due to different timing of collection and the influenced of season Jayaraman *et al.* (2003). In Badkhal lake Haryana the maximum temperature recorded was 31.2°C and minimum was 15.0°C (Kaushal and Sharma 2001). Sedemkar and Angadi (2003) recorded temperature in the range of 20.30 to 30.90°C in Jagat tank and Pala tank (1996). A positive correlation was observed between water temperature and dissolved oxygen. Shastri (2000) recorded water temperature ranges between 18 to 29°C. Surve *et al.* (2005) recorded the water temperature of Barul dam. It varied between 22.2 to 33.0°C, 22 to 32.9°C and 22.3 to 33.0°C at site I, II and III respectively.

Table 18.1: Monthly Mean Values of Temperature from Dhanegaon Reservoir Water Samples During the Year 2003–2004

Months	*Spot A*	*Spot B*	*Spot C*
June	24.6	24.8	25.0
July	21.8	22.2	21.7
August	20.3	21.4	20.8
September	21.1	22.3	22.4
October	23.2	23.4	23.5
November	25.4	25.4	25.5
December	21.2	21.6	21.8
January	21.7	21.8	21.9
February	22.9	23.0	22.7
March	23.5	23.7	23.8
April	28.1	28.5	28.9
May	30.0	30.4	30.3

All values are expressed in °C.

Table 18.2: Monthly Mean Values of Dissolved Oxygen from Dhanegaon Reservoir Water Samples During the Year 2003–2004

Months	*Spot A*	*Spot B*	*Spot C*
June	9.1	9.2	8.7
July	8.4	7.2	7.8
August	8.2	7.5	7.6
September	8.7	8.8	7.9
October	10.2	9.4	9.2
November	10.5	10.8	10.4
December	11.5	12.3	11.2
January	12.6	12.4	13.5
February	11.5	12.2	12.2
March	10.3	10.8	12.4
April	9.8	10.4	9.7
May	9.5	9.4	8.2

All values are expressed in mg/lit.

In the present investigation the amount of dissolved oxygen was found to be maximum in January in the year 2003-2004 with the values 12.6 mg/lit, 12.4 mg/lit and 13.5 mg/lit at spot A, B and C respectively. The amount of dissolved oxygen level were minimum in June at spot A 8.2 mg/lit, 7.2 mg/lit at spot B in June but 7.6 mg/lit at spot C in the month of May. Monthly mean values of dissolved oxygen are expressed in Table 18.2 and graphically represented in Figure 18.2. The seasonal fluctuation of dissolved oxygen is observed maximum in winter season and minimum in summer season. This is indicates the positive correlation of dissolved oxygen and water temperature. Increase the temperature

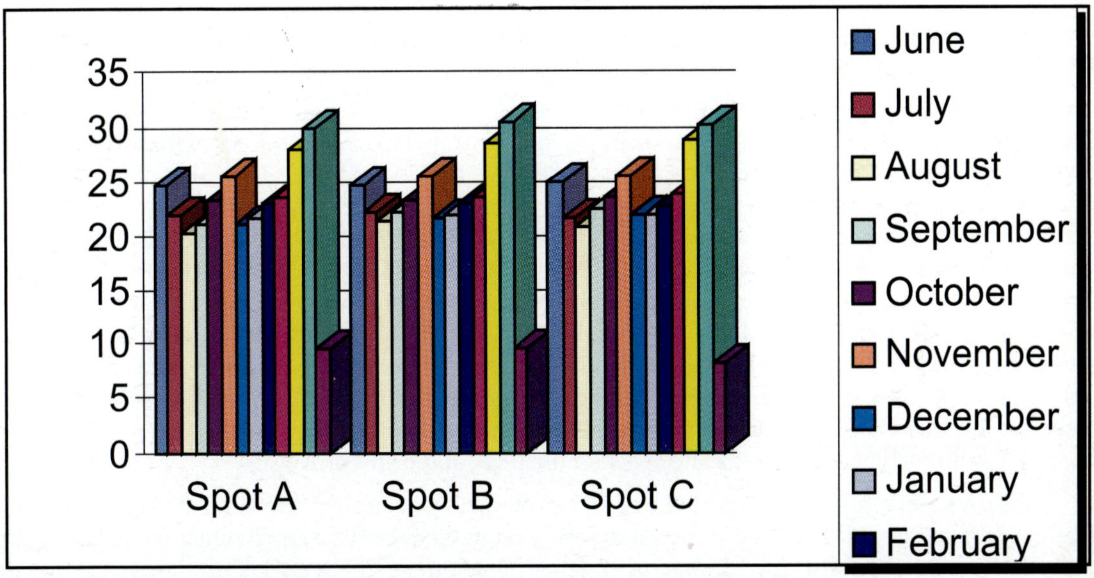

Figure 18.1: Monthly Variation in Water Temperature

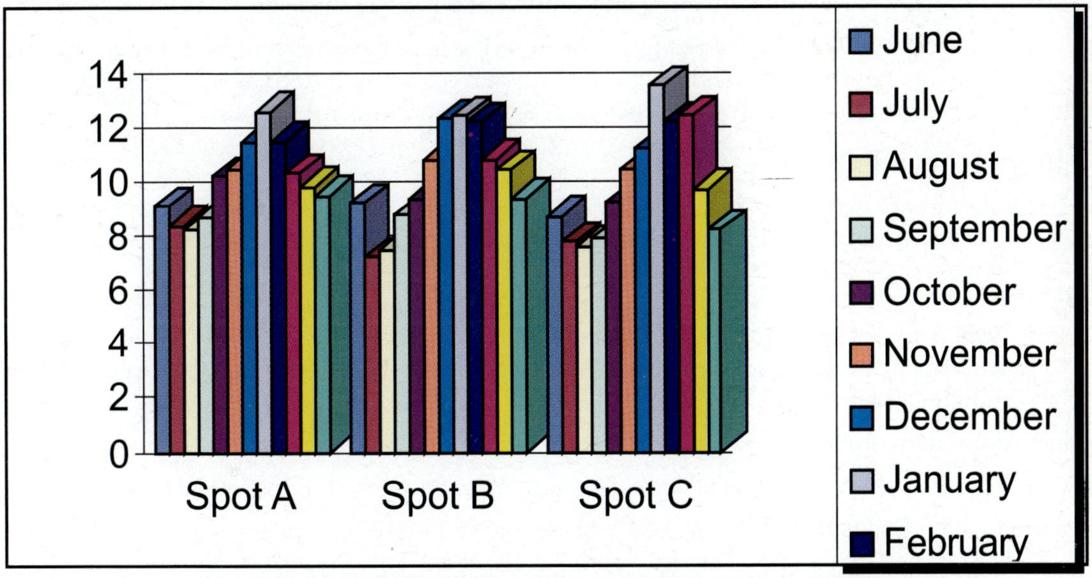

Figure 18.2: Monthly Variation of Dissolved Oxygen from Dhanegaon Reservoir Water Samples During the Year 2003–2004

of water in summer is as result the minimum amount of dissolved oxygen. Surve *et al.* (2005) reported that the dissolved oxygen fluctuated between 3.2 and 7.6 mg/lit, 3.4 to 7.9 mg/lit and 3.2 to 7.6 mg/lit at S1, S2 and S3 sites respectively.

The higher values of dissolved oxygen recorded in the winter season due to high solubility at low temperature and less degradation of organic substances in winter. The low oxygen in water can kill the fish and other organisms present in water. The potable range of drinking water of dissolved oxygen is above the 4.0 mg/lit. In the present investigation was observed the all value of dissolved oxygen within the permissible limit prescribed by WHO and ICMR.

The dissolved oxygen is one of the most important parameter in water quality and is an index of physical and biological processes going on in water. There are two main sources of dissolved oxygen in water, diffusion from air and photosynthetic activities within the water. Diffusion of oxygen from atmosphere is physical phenomenon and depends upon solubility of oxygen, which often effected by factors like temperature, water movements and salinity. Pulle *et al.* (2005) recorded dissolved oxygen was ranging from 3.08 to 4.28 mg/lit, 3.30 to 6.24 mg/lit and 4.78 to 6.28 mg/lit at three sites I, II and III respectively. The lowest value was observed during summer while highest value during winter months is similar finding in Dhanegaon reservoir. Sedamkar and Angadi (2003) recorded dissolved oxygen minimum 5.2 mg/lit in the month of January and maximum 10.4 mg/lit in the Pala tank near Gulbarga. A positive correlation was observed between water temperature and dissolved oxygen stated by Jha *et al.* (2003). The dissolved oxygen varied between 5.2 mg/lit and 12.6 mg/lit. It was increased from early may onwards and reached the peak at the end of August.

Raghuwanshi *et al.* (2005) recorded dissolved oxygen was ranging from 12.8 mg/lit to 18.6 mg/lit at lower lake Bhopal. The minimum value recorded in the month of April while maximum value recorded in the month of June. Radhika *et al.* (2004) studied on the Vallayani lake and reported that the dissolved oxygen in surface water ranging from 4.83 to 7.11 mg/lit, 4.11 to 6.81 mg/lit and 4.71 to 6.80 mg/lit during pre monsoon, monsoon and post monsoon respectively. No significant variations in dissolved oxygen were encounted both spatially and temporally in the water of Vallayani Lake.

In the present investigation the positive correlation between the water temperature and dissolved oxygen. The temperature is increases the amount of dissolved oxygen is decreases and the temperature is decreased the amount of dissolved oxygen increased in Dhanegaon reservoir.

Acknowledgements

The authors are thankful to Principal Dr. R. L. Kawale and Dr. D. G. Salunke, Head Department of Zoology and Fishery Science, Rajarshi Shahu College, Latur for providing necessary facilities.

References

APHA, 1995. *Standard Methods for Examination of Water and Wastewater*, 19th Edn. American Public Health Organization, American Water Works Association/Water Pollution Control Federation, Washington, D.C.

Indian Association of Aquatic Biologists, 1998. *Methodology of Water Analysis.*

ICMR, 1995. *Manual of Standard of Quality of Drinking Water Supplies*, 2nd Edn. Special Report Series 44.

Jayaraman, P.R., Gangadevi, T. and Vasudevan Nayer, T., 2000. Water quality studies on Karamana River Thiruvananthapuram District, South Kerala, India. *Poll. Res.*, 22(1): 89–100.

Jha, Prithviraj and Barel, Sudip, 2003. Hydrobiological study of lake Mirik in Darjelling Himalaya. *J. of Env. Biol.*, 4(3): 339–344.

Pulle, J.S., Khan, A.M., Ambore, N.E., Kadam, D.D. and Pawar, S.K., 2005. Assessment of groundwater quality of Nanded city. *Poll. Res.*, 24(3): 657–660.

Radhika, C.G., Mini, I. and Gangadevi, T., 2004. Studies on abiotic parameters of tropical freshwater lake–Vellayani lake, Thiruvananthapuram, Dist. Kerala. *Poll. Res.*, 23(1): 49–63.

Raghuwanshi, Arun K., 2005. The impact of physico-chemical parameter of lower lake Bophal on the productivity of *Eichhornia carassipes*. *Eco. Env and Cons.*, 11(3–4): 333–336.

Sedamkar Eswarlal and Angadi, S.B., 2003. Physico-chemical parameters of two freshwater bodied of Gulbarga, India with special reference to phytoplankton. *Poll. Res.*, 22(3): 411–422.

Surve, P.R., Ambore, N.E. and Pulle, J.S., 2005. Correlation co-efficients of some physico-chemical characteristics of Barul dam water, District Nanded (M.S.). *Poll. Res.*, 24(3): 653–656.

Chapter 19

Diurnal Changes of Some Physico-Chemical Factors in Thodga Reservoir of Latur District in Maharashtra

☆ *P.V. Patil and A.N. Kulkarni*

ABSTRACT

Thodga Dam is a medium reservoir with 165.00 hectors water spread area. The reservoir is constructed on Lendi River, one of the tributary of Manjara river system. The reservoir is on Ahmedpur–Thodga road 5 km away from Ahmedpur and 0.5 km from Thodga village. Fishermen's co-operative society uses this reservoir for fisheries activities. The water of this reservoir is also used for irrigation and drinking purpose. Diurnal changes were observed in the month of Nov-2002, changes in temperature, pH, Dissolved oxygen content, carbondioxiode content are estimated from 6 am to 6 am for 24 hours. Changes are discussed in the text.

Introduction

Reservoir constitutes an important inland fishery and also serves as an storage for the surface run of water various activities such as irrigation, power generation, fisheries etc. and as a means of providing employment to a sizable section of these water bodies, basic research on reservoir productivity is essential.This will help to increase production, potential and to set up target for fish production.

Studies on the physico-chemical changes are helpful to find out suitability of water for fish culture and pollution status for the reservoir. Diurnal changes in physico-chemical factors from Indian limonitic environments have been studied. Ganapati (1962), George (1966), Dobriyal and singh (1981), Sharma and Bhatt (1985), Panda *et al.* (1991), Sahu (1991), Shekar *et al.* (1993).

Many Aquatic organisms exhibit diurnal rhythms in their activities. The factors like, light, temperature and food are responsible for such activities.

Thodga reservoir is large minor irrigation tank, used for the fisheries activity. The details about the diurnal changes of this tank are not available, hence present work is undertaken to find out diurnal changes of some physico-chemical characters like temperature, pH, dissolved oxygen and CO_2 for 24 hours in the month of November 2002.

Material and Methods

For the estimation of diurnal changes of temperature pH, dissolved oxygen and CO_2 spot 'A' is selected near sluice gate. Changes were observed for 24 hours on date 2 to 3 November 2002 from 6.00 am to 6.00 am. For the estimation of dissolved oxygen and CO_2 methods described by Trivedi and Goel (1986) were used. Air and water temperature was recorded using the standard thermometer and pH is recorded using pH meter. Dissolve oxygen, free carbon dioxide was estimated at the spot only.

Results and Discussions

Diurnal changes in temperature, dissolved oxygen, Free CO_2 and pH are shown in Table 19.1 and Figure 19.1 Air temperature ranged from 22°C to 32°C. Air temperature increased from morning up to 2.00 pm. Then decreased onward.

Minimum water temperature is 18°C and maximum is 23°C. Temperature increased from morning up to 3.0 pm and it decreased afterwards.

Dissolved oxygen content ranged from 8.2 mg/lit to 9.0 mg/lit. It increased up to 4.0 pm and then it decreased later on.

Free carbon dioxide content ranged from 0.1 mg/lit to 1.0 mg/lit. Minimum carbon dioxide was observed in evening hours and maximum in morning hours, at 4.00 am.

pH ranged from 7.5 to 8.0 Philips (1927) observed diurnal fluctuations in hydrogen ion activity of Minnesota Lake.

Kato (1941) observed temperature increased from morning to noon and then it decreased during his studies on the freshwater region of the Palau tropical biological station II. George (1961) observed increased in temperature from morning to noon in the study of diurnal variations in two shallow ponds at Delhi.

Air temperature is more than water temperature. The same results were observed by Sahu *et al.* (1995), Sahu *et al.* (1996).

Hydrogen ion concentration was increased in afternoon hours, then it is decreased up to evening hours. George (1966) observed increased pH after 4.00 pm. Verduin (1959) demonstrated that nocturnal decrease in pH values was usually equal to the diurnal increase indicating that community respiration replaced the carbon dioxide absorption by photosynthesis.

Dissolved oxygen increased from morning 4.00 pm. It may be due to photosynthetic activities then it decreased afterwards to 5.00 pm. Sahu *et al.* (1995) and (1996) also observed same changes in dissolved oxygen content.

Carbon dioxide values were observed decreasing in the evening hours. This may be due to active process of photosynthesis. Generally CO_2 and oxygen showed inverse relationship.The factor affecting

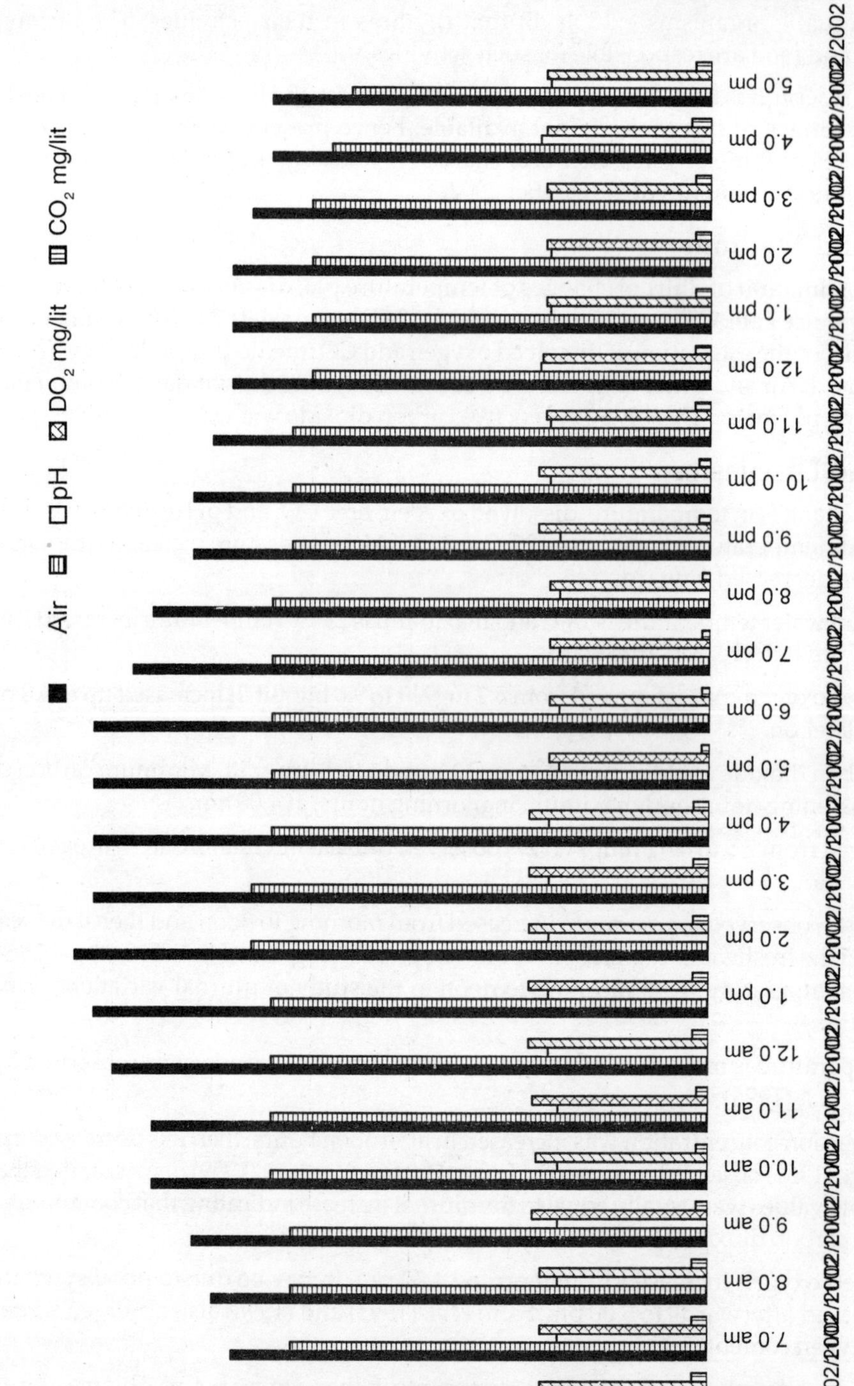

Figure 19.1: Diurnal Changes in Some Physio-chemical Factors of Thodga Reservoir on Date 02–03 November 2002

one of them should naturally be responsible for the other also. Thus increase in dissolved oxygen and decrease in free CO_2 values it may due to community respiration. Newwell (1957) also showed direct relationship between temperatures and dissolve oxygen in the water. Direct relationship exists between temperature and amount of oxygen dissolved in freshwater (Kato, 1941). In present study there is no direct relationship between dissolved oxygen content, water temperature and CO_2 content. This may be due to fluctuating cloudy weather and changes in the intensity of light during the observation.

Table 19.1: Diurnal Changes in Some Physico-chemical Factors of Thodga Reservoir on Date 02 to 03 November 2002

Date	*Time/hr*	*Temperature*		*pH*	*DO_2 mg/lit*	*CO_2 mg/lit*
		Air	*Water*			
02/11/2002	6.0 am	24	20	7.5	8.4	0.8
02/11/2002	7.0 am	24	21	7.5	8.4	0.8
02/11/2002	8.0 am	25	21	7.5	8.4	0.8
02/11/2002	9.0 am	26	21	7.5	8.8	0.6
02/11/2002	10.0 am	28	22	7.5	8.6	0.6
02/11/2002	11.0 am	28	22	7.5	8.8	0.6
02/11/2002	12.0 am	30	22	8.0	9.0	0.8
02/11/2002	1.0 pm	31	22	8.0	9.0	0.8
02/11/2002	2.0 pm	32	23	8.0	9.0	0.8
02/11/2002	3.0 pm	31	23	8.0	9.0	0.8
02/11/2002	4.0 pm	31	22	8.0	9.0	0.8
02/11/2002	5.0 pm	31	22	7.5	8.0	0.4
02/11/2002	6.0 pm	31	22	7.5	8.0	0.4
02/11/2002	7.0 pm	29	22	7.5	8.0	0.4
02/11/2002	8.0 pm	28	22	7.5	8.0	0.4
02/11/2002	9.0 pm	26	21	7.5	8.6	0.6
02/11/2002	10.0 pm	26	21	8.0	8.6	0.6
02/11/2002	11.0 pm	25	21	8.0	8.2	0.8
02/11/2002	12.0 pm	24	20	8.5	8.4	0.8
02/11/2002	1.0 pm	24	20	8.0	8.2	0.8
02/11/2002	2.0 pm	24	20	8.0	8.2	0.8
02/11/2002	3.0 pm	23	20	8.0	8.2	0.8
02/11/2002	4.0 pm	22	19	8.5	8.2	1.0
02/11/2002	5.0 pm	22	18	8.0	8.2	0.8

References

Dobriyal, A.K. and Singh, H.R., 1981. Diurnal variations in some aspects of limnology of the river mendakani from Garhwal Himalayas. *W.P.J. Zool.*, 3: 16–18.

George, M.G., 1961. Diurnal variation in two shallow ponds in Delhi India. *Hydrobiology*, 18: 256–257.

George Michel, R., 1966. Diurnal variations in physico-chemical factors and zooplankton in the surface layer of 3 freshwater fish pond. *Indian J. Fisheries*, 13(112): 49–82.

Ganapati, S.V., 1962. Diurnal variations on two shallow ponds in Delhi, India. *Hydrobiol.*, 18 (3): 265–273.

Panda, R.B., Sahu, B.K., Sinha, B.K. and Nayak, 1991. A comparative study and diurnal variations of physico-chemical characteristics of river, well and pond water at Rourkala Industrial complex of Orissa. *J. Ecotoxicol. Environ. Monit.*, 1(3): 206–217.

Kato, G., 1941. Studies on the freshwater regions in the compounds of the Palau tropical biological station II. Temperature, O_2 content and pH of the water. *Kagaku Nanyo (Sci of the South Sea)*, 3: 29–36.

Newell, B.S., 1957. The nocturnal reduction of dissolved oxygen in dams. *East African Agree. J.*, 23: 127–129.

Sharma, R.C. and Bhatt, 1985. Diurnal variations in physico-chemical and potamological parameters of the Bhagirathi river of Garhwal Himalaya. *Geobios*, 4: 104–108.

Sahu, B.K., Rao, R.J., Behara, S.K. and Pandit, R.K., 1991. A study of the aquatic pollution load in the river Brahmani. *Ph.D.Thesis*, Sambhalpur University, Sambhalpur.

Shaker, C., Malhotra, Y.R. and Datta, P.S, 1993. Diel variation in some abiotic parameters of water at Guptganga station of Torrential Neeru, Nullah, Bhadewal (Jammu). *J. Nat. Con.*, 5(2): 51–58.

Trivedi, R.K. and Goel, P.K., 1986. *Chemical and Biological Methods Water Pollution Studies*. Environmental Publications, Karad, India.

Verduin, J., 1959. Photosynthesis by aquatic communities in North Western. *Ohio. Ecol.*, 40: 377–383.

Chapter 20

Impact of Heavy Metals on Aquaculture and Fisheries: Its Determination in Water, Sediment, Fish and Feed Samples

☆ *P.H. Sapkale, V.B. Mulye and R.K. Sadawarte*

Introduction

Rapid Industrialization has led to metal pollution in coastal inland and marine waters. Heavy metals pollution is contributed by various ways such as agricultural runoffs, burning of fossil fuels, industrial effluents, domestic waste, transport, animal and human excretions, geological weathering, etc. in the aquatic environments. Heavy metals can affect the aquatic life as toxic substances in water, sediments or as a toxicant in food chain (Zyadah, 1995).

The sources of heavy metals include soil, effluents from industries such as metallurgy, mining, electroplating, pigment production, paints, tanneries, sewage sludge, agricultural chemicals and fertilizers, etc. In food processing industries, the contamination is due to the plants and equipments, catering operations, ceramic and enameled utensils, metal containers, etc.

Impact of Heavy Metals on Aquaculture and Fisheries

Heavy metals as such affect fisheries directly by exhibiting toxicity, and impairing metabolic activities and reproduction. Heavy metals affect fisheries by way of inhibiting the growth of fish food organisms and competing with nutrients. They also enter the food web resulting in biomagnifications. Contamination of heavy metals is also a major problem in sea foods, which has influence on the consumer health. The heavy metal toxicity in sea foods is mainly due to Hg, Cd, Pb, Cr, Se and As.

Mercury has persistence in the environment, bioaccumulation, and transport in the aquatic food chain, in fish mainly. Tissues and organs of fish have such a high affinity for mercury hat in can accumulate by a factor 9000, compared to the environment. In fish, mercury is in the form of methyl mercury, which is the most toxic form formed by intestinal bacteria, present in the slimes of the skin. Bacterial methylation of mercury is also been found in the organic fraction (sediment) of aquatic systems. The mercury content of Indian molluscan products was low and it was not exceeding 0.5 ppm, and in oysters, the highest levels are <0.3 ppm and Hg level in our fishery products, ranging between 0.03 to 0.336 ppm in shellfish. The first methyl mercury poisoning, minnamata disease, occurred in Japan, because of consumption of fish from water that was heavily contaminated by industrial wastewater containing mercury. In Iraq, the poisoning appeared to result from the ingestion of wheat treated with a mercurial fungicide.

In foods, only inorganic cadmium salts are present. Cadmium is found in cephalopods, mainly in squids and cuttlefishes. Most of the absorbed cadmium is retained in the kidneys in human beings. The half life of cadmium in human kidneys may be as long as 30 years. Cd poisoning first occurred in Japan, in 1947, which is the *'Itai-itai'* (ouch-ouch) disease, found among the inhabitants of fuchu area in Japan, who took rice, highly contaminated with Cd (300-2000 μg/day). Cd has been considered as a (human) carcinogen, primarily on the basis of induction of pulmonary tumors. It interferes with the metabolism of vitamin D, calcium and collage, leads to osteoporosis and osteomalacia in humans.

It is also a cumulative poison, which cause chronic kidney infection. Lead poisoning is known as "plumbism" Lead poisoning causes chronic kidney infection. Some of the shell fishes are known to have higher levels of Pb, especially the mussels. In foods, the Pb levels are due to indirect contamination, from packaging material and handling. The WHO (1987) has set the lowest observed adverse effect level (LOAEL) at 200 μg/l of blood. Permitted limit of lead in canned sea food is 5 ppm.

High concentration of copper in oysters leads to unpleasant metallic flavor and green colour. If the concentration of copper and zinc in water increases to 0.01 to 0.1 ppm respectively, oysters become green in colour, which is known as 'Green sick oyster'. Permitted limit of copper in canned sea food is 10ppm. Other metals such as arsenic, selenium and chromium have also been found in the aquatic animals of heavily polluted waters.

Determination of Heavy Metals in Aquatic Ecosystem

Atomic absorption spectroscopy (AA or AAS) is one of the commonest instrumental methods for analyzing for metals and some metalloids. Metalloids like antimony, arsenic, selenium, and tellurium are now routinely analyzed by hydride generation AAS. Determination of metals in the water or sediment or fish needs extraction of the same from the samples. Therefore, it is presented into three parts namely the instrumentation, preparation of samples and analysis of metals in the sample.

Instruments

Principle of AAS

Every element has a specific number of electrons, which are associated with the atomic nucleus in an orbital structure, which is unique to each element. The electrons occupy orbital positions in an orderly and predictable way. The lowest energy, most stable electronic configuration of an atom, known as the "ground state", is the normal orbital configuration for an atom (E_0). If energy of the right magnitude is applied to an atom, the energy will be absorbed by the atom, and an outer electron will be promoted to a less stable configuration (E_1, E_2, E_3 etc.) or "excited state" (Instruction manual Shimadzu corporation). As this state is unstable, the atom will immediately and spontaneously return to its ground state configuration. The electron will return to its initial, stable orbital position, and radiant energy equivalent to the amount of energy initially absorbed in the excitation process will be emitted.

Either the energy absorbed in the excitation process or the energy emitted in the decay process is measured and used for analytical purposes.

Atomic-absorption (AA) spectroscopy uses the absorption of light to measure the concentration of gas-phase atoms. Sine samples are usually liquids or solids, the analyte atoms or ions must be vaporized in a flame or graphite furnace. The atoms absorb ultraviolet or visible light and make transitions to higher electronic energy levels. The analyte concentration is determined from the amount of absorption. The relationship between absorbance and concentration of an absorbing species is explained by Beer-Lambert law (or Beer's law).

Components of AAS

Every absorption spectrometer must have components, which fulfill the three basic requirements. There must be: (A) a light source; (B) a sample compartment; and (C) a detector.

A. Light Sources

The two most common light sources used in atomic absorption are the "hollow cathode lamp" and the "electrodeless discharge lamp".

Hollow cathode lamp is a hollowed-out cylinder of the metal whose spectrum is to be produced (Instruction manual Shimadzu corporation). The anode and cathode are enclosed in a borosilicate or quartz cylinder normally filled with either neon or argon at low pressure. At the end of the glass cylinder is a window transparent to the emitted radiation. When an electrical potential is applied between the anode and cathode, some of the fill gas atoms are ionized. The positively charged fill gas ions accelerate through the electrical field to collide with the negatively charged cathode and dislodge individual metal atoms when excited. On return to ground state emit radiation.

For most elements, the hollow cathode lamp is a completely satisfactory source for atomic absorption. In a few cases, cases involve the more volatile elements where low intensity and short lamp life are a problem, where electrodeless discharge lamp is used. A small amount of the metal or salt of the element for which the source is to be used is sealed inside a quartz bulb. This bulb is placed inside a small, self-contained radiofrequency (RF) generator or "driver". When power is applied to the driver, a radiofrequency field is created. The coupled energy will vaporize and excite the atoms inside the bulb, causing them to emit their characteristic spectrum. EDL are available for the elements antimony, arsenic, bismuth, cadmium, cesium, germanium, lead, mercury phosphorus, potassium, rubidium, selenium, tellurium, thallium, tin and zinc.

B. Sample Compartment

AA spectroscopy requires that the analyte atoms be in the gas phase. Ions or atoms in a sample must undergo desolvation and vaporization in a high-temperature source such as a flame or graphite furnace. Flame AA can only analyze solutions, while graphite furnace AA can accept solutions, slurries, or solid samples. Flame AA uses a slot type burner to increase the path length, and therefore to increase the total absorbance. Sample solutions are usually aspirated with the gas flow into a nebulizing or mixing chamber to form small droplets before entering the flame. The graphite furnace has several advantages over a flame. It is a much more efficient atomizer than a flame and it can directly accept very small absolute quantities of sample. It also provides a reducing environment for easily oxidized elements. Samples are placed directly in the graphite furnace and the furnace is electrically heated in several steps to dry the sample, ash organic matter, and vaporize the analyte atoms.

(*i*) Flame Type atomizer

The sample cell, or atomizer, of the spectrometer must produce the ground state atoms necessary for atomic absorption to occur. This involves the application of thermal energy to break the bonds that hold atoms together as molecules. In premix type design, sample solution is aspirated through a nebulizer and sprayed as a fine aerosol into the mixing chamber. Here the sample aerosol is mixed with fuel and oxidant gases and carried to the burner head, which combustion and sample atomization occur. The finest droplets of sample mist, or aerosol, are carried with the combustion gases to the burner head, where atomization takes place. The fuel–oxidant mixture is selected on the basis of atomization temperature of the element to be determined.

(*ii*) Non-Flame Atomizers

(a) Graphite Furnace

Graphite furnace atomic absorption spectrometry (GFAAS) is also known by various other acronyms, including electrothermal atomic absorption spectrometry (ETAAS). In GFAAS, samples are deposited in a small graphite tube, which can then be heated to vaporize and atomize the analyte. Graphite tube furnace consists of a hollow graphite cylinder and surrounded by a metal jacket through which water is circulated and which is separated from the graphite tube by a gas space. An inert gas, usually argon is circulated in the gas space and enters the graphite tube openings in the cylinder wall. Detection limits for the graphite furnace fall in the ng/l range for most elements.

(b) Hydride Generation Methods

Respective hydride of unstable elements is produced to measure their content. Antimony, arsenic, bismuth, germanium, lead, selenium, tellurium and tin just some of the elements which, in trace amounts, have biological, environmental and technological importance.

(c) Light Separation and Detection

AA spectrometers use monochromators for light separation. The main purpose of the monochromator is to isolate the absorption line from background light due to interferences. Photomultiplier-tubes are the most common detectors for AA spectroscopy.

Preparation of Samples

(I) Water

Sampling and Sample Preservation

Quartz or TFE are the best sample containers but they are expensive so, polypropylene or borosilicate glass containers can be used. Soft glass containers should be avoided for samples containing metals in μg/l range. For collecting deepwater samples, samplers made of plastics should be used to avoid contamination. Samples used for analyzing silver should be stored in light absorbing containers. Sample should be preserved immediately after collection by acidifying with concentrated HNO_3 to pH < 2.0 and be stored in a refrigerator at 4°C. Samples used for the analysis of mercury should be preserved with 2 ml/l of 20 per cent $K_2Cr_2O_7$ solution (prepared in 1+1 HNO_3) and stored in a refrigerator not contaminated with mercury.

Digestion of Samples

Digestion of samples can be done in two ways *viz.*, wet digestion and dry ashing methods.

1. Wet Digestion

For analysis of total metals, sample is digested without filtration, for dissolved metals filtrate is digested or analyzed directly and for suspended metals, filter and material on it is digested. Filter blank is run simultaneously to obtain blank correction. Nitric acid will digest many samples adequately. Nitrate is an accepted matrix for both flame and electrothermal AAS. Other acids such as HCl, H_2SO_4, $HClO_4$ and HF can be used along with HNO_3 for specific samples. Completion of digestion is indicated by a light-coloured, clear solution. Sample should not be allowed to dry during digestion. Digested sample should be transferred to volumetric flask and made up to mark after cooling the digested samples. The sample is now ready for the analysis.

In recent years, use of microwave digestion is gaining momentum. Decomposition of water, sediment and biological samples with acids heated in microwave oven results in a rapid, accelerated digestion. Microwave radiation acts as a source of intense energy to rapidly heat the sample. Microwave heating is internal and external as opposed to conventional heating which is only external. Local internal heating takes place on individual particles can result in their rupture, thus exposing fresh surface to acid attack. This method also decreases the amount of acid needed and reduce or eliminate the loss of volatile elements.

2. Dry Ashing

Sample is taken in platinum or high-silica glass evaporating dish and evaporated to dryness. Then the dish is kept in muffle furnace at 400-450°C will we attain white ash. The ash is dissolved using minimum quantity of conc. HNO_3 and warm water. Then filtered and made to a known volume in the volumetric flask. The sample is now ready for the analysis.

(II) Sediment

The role of aquatic sediments as a sink for metal pollutant cannot be fully assessed by measuring the total metal concentration. In addition, determination of total element does not give accurate estimate of the likely environmental impact. Metal speciation occurring in the sediments is in turn expected to influence metal bioavailability, and thereby metal content in biota, in particular in the soft tissues of fish and mussels. Since availability critically depends upon the chemical form in which a metal is present in the sediment, considerable interest exists in trace element species. The sequential extraction scheme proposed by Tessier *et al.* (1979) can be used successfully to evaluate the properties of metals bound to different phases of sediment matrix. This procedure allows us to determine the distribution of metals in the following fractions: (1) loosely adsorbed to the surface of sediment particles; (2) bound to carbonates; (3) bound to iron and manganese oxides/hydroxides; (4) complexed by organic matter; and (5) incorporated into clay mineral lattices.

Sampling and Sample Processing

Van Veen/Ekman grab is used for the collection of sediment samples from deep water bodies. Sampler and the sample containers should be free from any metal contamination. The sample should be air-dried and ground using agate mortar to pass through a 74 µm stainless steel or nylon screen.

Digestion of Samples for Total Elements

Like water sample digestion, sediments also can be digested in two ways.

1. Acid digestion by Hot plate method
2. Fusion method

Decomposition of substances that are insoluble in acids can be done by using fusion agents (Na_2CO_3, Na_2O_2). Dissolution using acids is preferred to fusion because of the lower concentration of extraneous material in the final solution and less interference in the determination of elements if AAS or ICPES is used.

(III) Fish/Seed Samples

Sampling and Sample Preservation

After collecting the fish samples, they should be cleaned in deionised water and placed individually placed in plastic bags, labeled and transferred to freezer. Freezing and subsequent storage should be done at –18°C. The samples can be stored at this frozen condition for 6 months to one year. The samples (fish/feed) should be homogenized using a glass or quartz or Teflon pestle and mortar after removing lateral muscle and store in the freezer.

Digestion of Samples

Wet Digestion

1. Nitric acid–Perchloric acid method: Thawed and homogenized fish sample is wet digested in glass vessel with concentrated HNO_3, followed by concentrated $HClO_4$. After acid evaporation to dryness, the residue was dissolved in 5ml of 0.1 M HNO_3.
2. Nitric acid–Hydrogen peroxide method: Thawed and homogenized fish sample wet digested in glass beaker with freshly prepared 1:1 v/v HNO_3 and H_2O_2.

Dry Ashing

Thawed and homogenized fish sample is taken in silica crucible and dried 135-150°C for two hours. Then the sample is kept in a muffle furnace at 500°C for 1 hours (overnight). Nitric acid is added after cooling and kept on the hot plate for an hour Ash is dissolved using 1N HCl on hotplate and made to a known volume after it became cool.

Analysis

Based on the energy/temperature requirement of the element to be measured the fuel: oxidant type and ratio is selected for the flame AAS. Then the light source is selected and sample is aspirated on the flame and the absorbance is measured. From the standard absorbance, the concentration of the metal is question is calculated. Some metals like Hg, Cr may need special treatments, which may be followed to determine the accurate levels.

Conclusion

As heavy metal pollution causes havoc in aquatic environment in general and public health in particular, any management measure should address at the source level rather than at the sink. Accurate and appropriate analytical methods should be followed to initiate successful management strategy.

References

AAS Cookbook–Section–1, Basic conditions of Analysis of Atomic Absorption Spectrophotometry in AA–6800, Instruction Manual, Shimadzu Corporation (Asia Pacific) Pte Ltd., Singapore.

American Public Health Association, 1985. *Standard Methods for the Examination of Water and Wastewater*, 16th Edn. APHA, Washington, DC.

Beaty, R.D. and Kerber, J.D., 1993. *Concepts, Instrumentation and Techniques in Atomic Absorption Spectrophotometry*, 2nd Edn. The Perkin–Elmer Corporation, Norwalk, CT, USA.

Sparks *et al.* (Ed), 1996. *Methods of Soil Analysis, Part III: Chemical Methods*. SSSA Book Series 5, SSSA and ASA, Madison, WI.

Tessier, A., Cambell, P.G.C. and Bission, M., 1979. Sequential extraction procedure for the speciation of particulate trace metals. *Anal. Chem.*, 51: 844–851.

Vogel's *Textbook of Quantitative Analysis*. Longman Publication.

Zyadah, M., 1995. Environmental impact assessment of pollution in lake Manzalah and its effect on fishes. *Ph.D. Thesis*, Faculty of Science, Mansura University, Egypt, 127.

Chapter 21

Effect of Dimethoate on Blood Sugar Level of Freshwater Fish, *Macronus vittatus*

☆ *D.S. Rathod, M.V. Lokhande and V.S. Shembekar*

ABSTRACT

Pesticide, dimethoate as pollutant, was used to study its effect on blood sugar level in the freshwater fish, *Macronus vittatus*. The fish *Macronus vittatus* was exposed to lethal (5.78ppm) and sublethal (0.578ppm) concentration of dimethoate for 24 hours. Blood sugar level/100 ml of blood was determined at 6,12,18 and 24 hours interval. It was observed that at 6 and 12 hours of exposure period blood sugar level increased where as from 12 hour onwards the blood sugar level was found to be decreased in both lethal and sublethal concentration. Increase in the blood sugar level was more in lethal concentration than in sublethal concentration at same time intervals (6 to 12 hours). After 12 hours of exposure, the decrease in blood sugar level was also more in lethal concentration when compared to sublethal concentration.

From these results it can be inferred that under diemethoate stress, blood sugar level initially increases in both lethal and sublethal concentration. However, with increase in exposure time it decrease probably due to poisoning of the metabolic process.

Introduction

Growing human population has created a need for increased food production hence in the recent years pesticides are progressively used for the protection of crops from pests. The use of pesticides has undoubtedly increased agricultural production but on the other side, the dark part of the pesticides is highly terrific. Pesticides percolate through soil and reach water bodies through run-off rain water. They contaminate the water bodies and endanger the aquatic life.

In aquatic environment fishes are the most important group not only because of their amenity and nutritive value, but also because of their mobility and sensitivity towards change occurring in their surroundings. Hence, fish serves as a biological indicator to detect the degree of pollution in water bodies.

The study of toxicity of pesticide is much important to avoid its impact on non-target organism like fishes. Many pesticides are found to affect the central nervous system and metabolic activities either directly or indirectly.

Metabolic parameters like ratio of oxygen consumption of blood sugar level and 'disruption of energy' balance have been studied by several workers to detect the metabolic distress due to sublethal effect of pollutants (Bayne, 1973; Gabbott and Bayne, 1973, Thompson and Bayne, 1974; and Bayne *et al.*, 1978).

Pesticide alter body components to variable degree depending on the concentration and exposure time. Therefore, it becomes necessary to evaluate the nature and extent of alteration in biochemical components of fish.

The present investigation was carried out to see the effect of an organophosphate, dimethoate on blood sugar level of freshwater fish *Macronus vittatus.*

Materials and Methods

The freshwater fish, *Macronus vittatus* were collected from Manjara river, Latur and brought to laboratory. These fishes were observed for any pathological symptoms and then placed in 0.1 per cent solution of potassium permagnate ($KMnO_4$) for 2 minutes so as to avoid any dermal infection. The fish were then washed with water and acclimatized to laboratory conditions for two weeks in glass aquaria. The physico-chemical parameters of water were analysed by following standard method suggested by APHA (1989) and IAAB (1998).

During acclimatization the fish were provided with a diet consisting of live earthworms. Food supply was withdrown 24 hours prior to the experimentation. A commercial grade of pesticide, Dimethoate (Roger)-30 per cent was used for bioassay test. A stock solution of the toxicant was prepared and then few concentrations from stock solutions were prepared as per the dilution technique (APHA 1989).

For experimentation, laboratory acclimatized fishes were divided into three groups of 10 fish per aquarium. Group 'A' served as control. Group 'B' and 'C' were exposed to lethal and sublethal concentration of dimethoate. The blood sample was collected from control and exposed fishes after 6, 12, 18 and 24 hours interval and the blood sugar was estimated by the method of Kemp and Mayers (1954).

Results and Discussion

It was observed that blood sugar level/100 ml of blood after dimethoate treatment was increased when compared to control. The value for control group was 3.5±1.25 mg. After 6 hours exposure in sublethal concentration (0.578 ppm). It was increased to 24.0±2.85 mg and in lethal concentration it significantly increased to 28.0±4.21 mg. 12 hours exposure in dimethoate brought the blood sugar level to 62.0±3.78 mg. in sublethal concentration and to 68.0±3.92 mg. in lethal concentration. However, 18 hours and 24 hours exposure decreased the blood sugar level. The effect of dimethoate on blood sugar level of *Macronus vittatus* is represented in Table 21.2.

Table 21.1: Physico-chemical Parameters of Water

Sl.No.	Parameters	Value
1.	Dissolved Oxygen (mg/l)	7.0 ± 1.0
2.	pH	7.4 ± 0.2
3.	Temperature (°C)	28.0 ± 4
4.	Total hardness $CaCO_3$ (Mg/l)	170 ± 8.0

Table 21.2: Effect of Dimethoate on Blood Sugar Level of *Macronus vittatus*

Exposure Time	Control (Group A)	Experimental	
		Group B (Sub lethal Conc. i.e. 0.578 ppm)	Group C (Lethal Conc. i.e. 5.78 ppm)
6 hours	3.5 ± 1.25 mg	24.0 ± 2.85 mg	28.0 ± 4.21 mg
12 hours	–	62.0 ± 3.78 mg	68.0 ± 3.92 mg
18 hours	–	14.85 ± 3.85 mg	18.89 ± 8.3 mg
24 hours	–	11.0 ± 1.2 mg	13.25 ± 1.90 mg

Dimethoate treatment in freshwater fish, *Macronus vittatus* caused rapid hyperglycemia upto 12 hours causing relatively more increase in blood sugar level. This hyperglycemic response during minimum exposure time suggests and attempt for homeostasis. The prolonged exposure (18 to 24 hours) exhibited decrease in blood sugar level, which suggest poisoning of the metabolic process and hence depressed activity of homeostatic mechanism resulting in the inability to acclimate to the altered environment. Thomas *et al.* (1981), reported that depending upon degree of stress the hyperelycemia turn to hypoglycemia. Mirajkar *et al.*(1984) worked on effect of dimecron on blood sugar level in the freshwater prawn, *Macrobrachium kistnenis* and observed that the blood sugar level increased as compared to the control.

Bhatia *et al.* (1973) worked on effect of dieldrin on hepatic carbohydrate metabolism and protein biosynthesis in vivo, fish and rat reported that decrease in glycogen and increase in blood glucose level under pesticidal toxicity. Parvathi (1982)–Worked on physiological and biochemical responses of carps, *Cyprinus carpio, Labeo rohita, Cirrihina mrigala* subjectd malathion exposure to different temperature that, the decrease in glycogen and increase in glucose. Ghosh (1987)–Worked on toxic impact of the three organophosphate pesticides on carbohydrate metabolism in freshwater Indian catfish *Clarius batrachus*, reported that decrease in the level of glycogen in tissue like liver, gills and brain under treatment lihocin and suggested that, the stored glycogen is hydrolysed into glucose which utilized as a source of energy. Abdul Naveed *et al.* (2006)–Worked on toxicity of lihocin of on the activities of glycolytic and glycongenic enzymes of fish, *Channa punctatus* observed that the level of glycogen and pyruvate decreased, while glucose and the lactic acid level increased.

Conclusion

In case of *Macronus vittatus* the release of hyperglycemic hormone due to the dimethoate stress might have resulted in an increase in blood sugar level or it might be possible that the fish while adjusting to the new environment had to undergo stress and this probably caused increase in blood sugar. The increase in blood glucose level may be due to decrease in glycogen synthesizing potentiality of the tissues as a consequence of cellular damage.

References

APHA, 1989. *Standard Methods for Examination of Water and Wastewater.* American Public Health Association, 20th Edn. Washington D.C.

Bayne, B.L., 1973. Physiological changes in *Mytilus edulis* induced by temperature and nutritive stress. *J. Mar. Biol. Assoc.* U.K. 53: 39–58 pp.

Bayne, B.N., 1976. Watch of mussel. *Mar. Poll. Bull.*, 7: 227.

Bayne, B.L., M.M., Davey, J.T. and Scullar, G., 1978. Physiological responses of *Mytilus edulis* L to parasitic infestation by *Mytilicota intestinals. J. du counclint. Explor. Mer.*, 33: 12–70.

Bhatia, S.C., Sharma, S.C. and Venkita Subramanian, T.A., 1973. Effect of dieldrin on hepatic carbohydrate metabolism and protein biosynthesis *in vivo. Toxical. Appl. Pharmacol*, 24: 216.

Gabbot, P.A. and Bayne, B.L., 1973. Biochemical effect of temperature and nutritive stress on *Mytilus edulis* L. *J. Mar. Biol. Ass.*, U.K., 53: 269–286.

Garnett, J. and Leeling, N.C., *Ann. Entomol. Soc. Amer.*, 65: 299.

Ghosh, 1987. Toxic impact of three organophosphate pesticide on carbohydrate metabolism in a freshwater Indian cat fish, *Clarias batrachus. Proc. Indian, Nat. Sci. Acad., B.* 53(2): 135–142.

IAAB, 1998. *Methodology for Water Analysis*, Hyderabad.

Kemp, M.B.A. and Mayers, D.K., 1954. A colorometric mocromethod for the determination of glucose, *Biochem, J.*, 56: 639–645.

Mirajkar, M.S., Khan, A.K. and Sarojini, R., 1984. Effect of dimecron on blood sugar level in the freshwater prawn. *Macrobranchium kistensis,* 101–104 pp.

Parvathi, 1982. Physiological and biological responses of *Carp, Cyprinus caripo, Labeo rohita* and *Cirrihina mrigala* subjected to melathion exposure to different temperatures. *Ph.D. Thesis*, S.K. University, Anantpur, India.

Samarnayaka, M., 1974. *Gen. Comp Endocrinal.*, 24: 424.

Thomas, P., Carr, R.S. and Neff, J.M., 1981. In: *Biological Monitoring of Marine Pollutants,* (Eds.) F.J. Vernberg, A. Calabress, E.P. Thrunberg and W.B. Varnberg. Academic Press INC (London) Ltd. New York, 10003 pp.

Thompson, R.J. and Bayne, B.J., 1974. Active metabolism associated with feeding in the *Mytilus edulis* L. *J. Expt. Mas Biol. and Ecol.*, 8: 191–212.

Index